Modern Birkhäuser Classics

Many of the original research and survey monographs, as well as textbooks, in pure and applied mathematics published by Birkhäuser in recent decades have been groundbreaking and have come to be regarded as foundational to the subject. Through the MBC Series, a select number of these modern classics, entirely uncorrected, are being re-released in paperback (and as eBooks) to ensure that these treasures remain accessible to new generations of students, scholars, and researchers.

More information about this series at http://www.springer.com/series/7588

Pierre-Alain Cherix · Michael Cowling
Paul Jolissaint · Pierre Julg
Alain Valette

Groups with the Haagerup Property

Gromov's a-T-menability

Reprint of the 2001 Edition

 Birkhäuser

Pierre-Alain Cherix
Section de Mathématiques
Université de Genève
Genéva
Switzerland

Pierre Julg
Bâtiment de Mathématiques, MAPMO
Université d'Orléans
Orléans Cedex 2
France

Michael Cowling
School of Mathematics and Statistics,
   The Red Centre
University of New South Wales
Sydney, NSW
Australia

Alain Valette
Institut de Mathématiques
Université de Neuchâtel
Neuchâtel
Switzerland

Paul Jolissaint
Institut de Mathématiques
Université de Neuchâtel
Neuchâtel
Switzerland

ISSN 2197-1803   ISSN 2197-1811   (electronic)
Modern Birkhäuser Classics
ISBN 978-3-0348-0905-4   ISBN 978-3-0348-0906-1   (eBook)
DOI 10.1007/978-3-0348-0906-1

Library of Congress Control Number: 2015931242

Mathematics Subject Classification (2010): 20-XX, 22Dxx, 22Exx, 43-XX, 51-XX, 22-02, 22D10, 22D25, 22E30, 43A07, 43A65, 46Lxx

Springer Basel Heidelberg New York Dordrecht London
© Springer Basel 2001
Reprint of the 1st edition 2001 by Birkhäuser Verlag, Switzerland
Originally published as volume 197 in the Progress in Mathematics series

Cover design: deblik, Berlin

Printed on acid-free paper

Springer Basel AG is part of Springer Science+Business Media (www.birkhauser-science.com)

Contents

Chapter 1

Introduction
by Alain Valette

1.1 Basic definitions

1.1.1 The Haagerup property, or a-T-menability

For a second countable, locally compact group G, consider the following four properties:

(1) there exists a continuous function $\psi\colon G \to \mathbb{R}^+$ which is conditionally negative definite and proper, that is, $\lim_{g\to\infty} \psi(g) = \infty$;

(2) G has the **Haagerup approximation property**, in the sense of C.A. Akemann and M. Walter [AW81] and M. Choda [Cho83], or **property** C_0 in the sense of V. Bergelson and J. Rosenblatt [BR88]: there exists a sequence $(\phi_n)_{n\in\mathbb{N}}$ of continuous, normalized (i.e., $\phi_n(1) = 1$) positive definite functions on G, vanishing at infinity on G, and converging to 1 uniformly on compact subsets of G;

(3) G is **a-T-menable**, as M. Gromov meant it in 1986 ([Gro88, 4.5.C]): there exists a (strongly continuous, unitary) representation of G, weakly containing the trivial representation, whose matrix coefficients vanish at infinity on G (a representation with matrix coefficients vanishing at infinity will be called a C_0-**representation**);

(4) G is **a-T-menable**, as Gromov meant it in 1992 ([Gro93, 7.A and 7.E]): there exists a continuous, isometric action α of G on some affine Hilbert space $\mathcal{H}$, which is metrically proper (that is, for all bounded subsets B of $\mathcal{H}$, the set $\{g \in G : \alpha(g)B \cap B \neq \emptyset\}$ is relatively compact in G).

It was gradually realized that these conditions are actually equivalent, hence define a unique class of groups (see [AW81] for the equivalence of (1) and

© Springer Basel 2001
P.-A. Cherix et al., *Groups with the Haagerup Property*,
Modern Birkhäuser Classics, DOI 10.1007/978-3-0348-0906-1_1

(2), [Jol00] for the equivalence of (1) and (3), and [BCV95] for the equivalence of (1) and (4), proved there for discrete groups, but the proof goes over without change to the general case). We believe that Gromov was already aware in 1992 of the equivalence of (3) and (4), but that he had no formal proof[1].

Definition 1.1.1. A second countable, locally compact group has the **Haagerup property** if it satisfies one and hence all of the equivalent conditions (1) to (4) above.

A short proof of the equivalence of conditions (1) to (4) will be given in Theorem 2.1.1 below.

This set of cognate papers is devoted to the study of the class of groups with the Haagerup property.

1.1.2 Kazhdan's property (T)

It is patently obvious that each of the conditions (1) to (4) above is designed as a strong negation to **Kazhdan's property (T)**; indeed, here are the four corresponding equivalent formulations of property (T) for G (see [HV89] for the proofs of the equivalences, and for many examples of groups with property (T)):

(1) every continuous, conditionally negative definite function on G is bounded;

(2) whenever a sequence of continuous, normalized, positive definite functions on G converges to 1 uniformly on compact subsets of G, then it converges to 1 uniformly on G;

(3) if a representation of G contains the trivial representation weakly, then it contains it strongly (that is, the representation has G-fixed vectors);

(4) every continuous, isometric action of G on an affine Hilbert space has a fixed point.

1.2 Examples

We give here a list of examples (roughly in chronological order) of groups with the Haagerup property. They illustrate how large this class of groups is. It will be noticed that most of these groups are of geometric origin. As advocated by Julg [Jul98] and Y.A. Neretin [Ner98], we will describe, whenever possible, an explicit proper isometric action on some affine Hilbert space.

---

[1] Otherwise, Gromov would not have asked, as he did in [Gro93, 7.E], whether amenable groups are a-T-menable; indeed, given characterization (3), this fact is obvious, as observed in [Jol00].

1.2.1 Compact groups

Compact groups have the Haagerup property, trivially. They also have property (T). Conversely, it is clear from the conditions above that a group with both the Haagerup property and property (T) has to be compact. As a consequence, any continuous homomorphism from a group with property (T) to a group with the Haagerup property has relatively compact image (this is the guiding principle in [HV89, Chap. 6]).

1.2.2 $SO(n, 1)$ and $SU(n, 1)$

The Lie groups $SO(n, 1)$ and $SU(n, 1)$ (the isometry groups of the n-dimensional real and complex hyperbolic spaces respectively) have the Haagerup property. Proper affine isometric actions were constructed by A.M. Vershik, I.M. Gel'fand and M.I. Graev [VGG73], [VGG74] (see also [Ner98, 1.1–1.3], as well as Chapter 3 below). On the other hand, it was proved by J. Faraut and K. Harzallah [FH74] (see also [HV89, 6p. 79]) that, denoting by d the hyperbolic distance and by x_0 any point in real or complex hyperbolic space, the function $g \mapsto d(gx_0, x_0)$ is conditionally negative definite on $SO(n, 1)$ and on $SU(n, 1)$; for $SO(n, 1)$, A.G. Robertson ([Rob98, Cor. 2.5]) constructed the associated affine action on the L^2 space of the space of half-spaces of real hyperbolic space. Note that Y. Shalom [Sha99] has proved that, if α is an affine isometric action of $SO(n, 1)$ or $SU(n, 1)$, then either α is proper or α has a fixed point.

1.2.3 Groups acting properly on trees

Free groups on a finite set of generators have the Haagerup property; indeed, U. Haagerup [Haa79] established the seminal result that the word length with respect to a free generating subset is a conditionally negative definite function on the group. (Since the Haagerup property is clearly inherited by closed subgroups, this means that the free group on countably many generators also has the Haagerup property.) Haagerup's result was reinterpreted in terms of group actions on trees by several people, see [Wat81], [Alp82], [JV84, Lem. 2.3] (see also [HV89, p. 69]), [Mar91, Prop. 3.6], [Pav91]. Denote the distance function on a tree X by d, and an arbitrary vertex in X by x_0; then the function $g \mapsto d(gx_0, x_0)$ is conditionally negative definite on the automorphism group of X. The corresponding affine isometric action on the ℓ^2 space of the set of oriented edges of the tree is implicit in most of the references above[2]. In particular, any group acting properly on a locally finite tree, has the Haagerup property.

---

[2]This affine action is used in [Sha99] to prove a superrigidity result for actions of lattices on trees. For groups acting on homogeneous, locally finite trees, a different affine isometric action is constructed in [Ner98, 1.4].

1.2.4   Groups acting properly on $\mathbb{R}$-trees

The case of trees may be generalized to $\mathbb{R}$-trees: the function $g \mapsto d(gx_0, x_0)$ is conditionally negative definite on the isometry group of a $\mathbb{R}$-tree (see [Boż89], [HV89, pp. 73–74]). The corresponding affine isometric action (on the L^2 space of the $\mathbb{R}$-tree, for a suitably defined measure) appears in [Val90] and is thoroughly studied in [Ner98]. In particular, groups acting metrically properly on a $\mathbb{R}$-tree have the Haagerup property.

1.2.5   Coxeter groups

Coxeter groups have the Haagerup property; more precisely, M. Bożejko, T. Januszkiewicz and R. Spatzier [BJS88] (see also [HV89, p. 76]) proved that, for a Coxeter system (W, S), the word length with respect to S is conditionally negative definite on W. The corresponding affine isometric action is on the ℓ^2 space of the set of roots of the associated Coxeter complex.

1.2.6   Amenable groups

Amenable groups have the Haagerup property; this was proved in [BCV95] by using Følner sets to produce a proper affine isometric action on a Hilbert space, thereby answering a question of Gromov ([Gro93, 7.A and 7.E]). As observed in [Jol00], given the equivalence of (3) and (4), the Haagerup property for amenable groups becomes essentially trivial: indeed, for any group G, the left regular representation λ_G is a C_0-representation, and if G is amenable, then λ_G weakly contains the trivial representation.

1.2.7   Groups acting on spaces with walls

It was observed by G. Niblo and L. Reeves [NR97] that, if X is a CAT(0) cubical complex[3], then for any vertex x_0 of X, the function $g \mapsto d_c(gx_0, x_0)$ is conditionally negative definite on the automorphism group of X; here d_c denotes the **combinatorial** distance on X, this is, the distance on the 1-skeleton. Indeed, there is a well-defined notion of walls separating X (see [Sag95]), and the corresponding affine action on the ℓ^2 space of the set of half-spaces is implicit. This was generalized by F. Haglund and F. Paulin [HP98] to even polyhedral complexes (also called zonotopal complexes) which are CAT(0). As a consequence, a group acting properly on a CAT(0) even polygonal complex (in particular on a CAT(0) cubical complex) has the Haagerup property. As noticed in [Bri], the class of groups acting properly on CAT(0) cubical complexes is fairly large; it contains in particular the finitely presented, simple,

---

[3]CAT is derived from the surnames of É. Cartan, A.D. Alexandrov and A. Toponogov.

torsion-free groups constructed by M. Burger and S. Mozes [BM97]. In his thesis [Far00], D.S. Farley showed that this class also contains Thompson's group F.

An analogous situation was studied by W. Ballmann and J. Świątkowski [BŚ97]: a 2-dimensional polygonal complex X is a (k, l)-**complex** if each face has at least k edges and each link of vertex has girth at least l. Assume that X is either a $(6, 3)$- or a $(4, 4)$-complex, and is simply connected. If d is the metric on X that turns X into a Hadamard space, then by [BŚ97, Lemmata 5.5 and 5.6], there exists a conditionally negative definite kernel N on the set X_2 of faces of X, and N is bi-Lipschitz equivalent to d in the following sense: if $\tilde{p}$ denotes the barycentre of the face $p \in X_2$, there exist positive constants C_1 and C_2 such that

$$C_1\, d(\tilde{p}, \tilde{q}) \le N(p, q) \le C_2\, d(\tilde{p}, \tilde{q})$$

for all $p, q \in X_2$. Hence, for a given $p_0 \in X_2$, the function $g \mapsto N(gp_0, p_0)$ is conditionally negative definite on the automorphism group of X. In particular a group acting properly on a simply connected, $(6, 3)$- or $(4, 4)$-complex, has the Haagerup property.

In [HP98], Haglund and Paulin described a structure of space with walls, which underlies all the examples of this type, as well as Examples 1.2.3 and 1.2.5. A **space with walls** is a set X endowed with a (nonempty) family $\mathcal{W}$ of partitions of X into 2 classes, called walls, such that the number $w(x, y)$ of walls separating any two distinct points x and y in X is finite. A group G acts properly on a space with walls $(X, \mathcal{W})$ if G acts on X, preserving the family $\mathcal{W}$, and for some $x_0 \in X$ (hence for all $x_0 \in X$), the function $g \mapsto w(gx_0, x_0)$ is proper on G. It turns out that this function is conditionally negative definite on G, so that such a G has the Haagerup property. A proof of this unpublished result of Haglund, Paulin and the author of this chapter may be found in [Sha99]; see also Corollary 7.4.2 below.

One of the purposes of this book is to provide new examples of groups with the Haagerup property.

1.3 What is the Haagerup property good for?

Property (T) is often considered as a representation-theoretic form of rigidity. By way of contrast, it may be said that groups with the Haagerup property are *strongly nonrigid*. To explain what we mean, suppose that Γ is a discrete group with the Haagerup property, and let ψ be a proper, conditionally negative definite function on Γ; by perturbing ψ if necessary by a bounded function, we may assume that $\psi(g) = 0$ if and only if $g = 1$ (see Lemma 6.2.1 below). By Schoenberg's theorem (see, for instance, [HV89, p. 66]), when $t \geq 0$, the function $e^{-t\psi}$ is positive definite on Γ. Let π_t be the unitary representation associated with $e^{-t\psi}$ by the Gel'fand–Naimark–Segal construction. Then $(\pi_t)_{0 \leq t \leq \infty}$ is a one-parameter family of unitary representations of Γ that "interpolates" between the trivial representation (at $t = 0$) and the regular representation (at $t = \infty$).

1.3.1 Harmonic analysis: weak amenability

The Haagerup property appeared in connection with approximation properties of operator algebras (see, for instance, Haagerup [Haa79], Choda [Cho83], C. Anantharaman-Delaroche [AD95], Robertson [Rob93]). For example, Robertson [Rob93, Theorem C] proves that, if Γ_1 is a discrete group with the Haagerup property and Γ_2 is a countable subgroup with property (T) of the unitary group $\mathcal{U}(W^*\Gamma_1)$ of the von Neumann algebra of Γ_1, then Γ_2 is relatively compact for the L^2 norm topology on $\mathcal{U}(W^*\Gamma_1)$; in particular Γ_2 is residually finite.

Applications of the Haagerup property to harmonic analysis appear in [Haa79] and [JV91]; they are especially useful when combined with property (RD) (also called the Haagerup inequality): for example, if a locally compact group G satisfies property (RD) with respect to a length function L which is conditionally negative definite (in particular, G has the Haagerup property, since L is proper), then there is a simple characterization of positive definite functions ϕ weakly associated with the regular representation of G: the function ϕ is the limit of compactly supported positive definite functions, uniformly on compact subsets of G if and only if $\phi e^{-tL} \in L^2(G)$ for all $t > 0$ ([JV91, Thm 3]); if the Haar measure of the balls $\{g \in G : L(g) \leq s\}$ grows at most exponentially with s, this is equivalent to $\phi \in L^{2+\epsilon}(G)$ for all $\epsilon > 0$ ([JV91, p. 811]). A different kind of application to harmonic analysis was obtained by Bergelson and Rosenblatt ([BR88, Thm 2.5]): for a separable infinite-dimensional Hilbert space $\mathcal{H}$, denote by $\text{Rep}(G, \mathcal{H})$ the set of unitary representations of G in $\mathcal{H}$, endowed with the topology of uniform convergence on compact subsets; if G is noncompact and has the Haagerup property, then the set of representations without finite-dimensional subrepresentations is a dense G_δ in $\text{Rep}(G, \mathcal{H})$; in fact, the subset of C_0-representations is already dense in $\text{Rep}(G, \mathcal{H})$[4].

Since every amenable group has the Haagerup property, the latter can be seen as a weak form of amenability. As such, it is interesting to compare it with other weak forms of amenability, especially that introduced by Cowl-

---

[4]A finite-dimensional unitary representation π cannot be C_0, since $|\det \pi(g)| = 1$ for every $g \in G$; for that reason, a C_0-representation has no finite-dimensional subrepresentations.

ing and Haagerup [CH89]. Let $A(G)$ be the Fourier algebra [Eym64] of the locally compact group G, this is, the space of matrix coefficients of the left regular representation of G, endowed with pointwise multiplication. It is a well-known result of H. Leptin [Lep68] that G is amenable if and only if $A(G)$ admits an approximate unit (that is, a net $(u_i)_{i \in I}$ converging to the constant function 1 uniformly on compact subsets of G) with norm bounded by 1 in $A(G)$. Now let $M_0 A(G)$ be the algebra of completely bounded multipliers of $A(G)$; the inclusion of $A(G)$ in $M_0 A(G)$ is norm decreasing. We say that G **is weakly amenable with Cowling–Haagerup constant** 1 if there exists an approximate unit $(u_i)_{i \in I}$ in $A(G)$ such that $\|u_i\|_{M_0 A(G)} \leq 1$ for all $i \in I$. Amenable groups, Lie groups locally isomorphic to $SO(n, 1)$ or $SU(n, 1)$ (see [DCH85] for $SO(n, 1)$, [Cow83] for $SU(n, 1)$, and [LH90] for $\widetilde{SU}(n, 1)$), groups acting properly on trees (see [Szw91] and [Val90]), and Coxeter groups (see [Jan98], building on previous work in [Val93]), and closed subgroups thereof, are all examples of groups which are weakly amenable with Cowling–Haagerup constant 1. Comparing this list with that in Section 1.2 above, we see that all these groups also have the Haagerup property. Michael Cowling conjectures that a group has the Haagerup property if and only if it is weakly amenable with Cowling–Haagerup constant 1.

1.3.2 K-amenability

Another weak form of amenability is K-theoretic amenability (or K-amenability for short), introduced by J. Cuntz [Cun83]. To formulate it, we need two C*-algebraic definitions. For a locally compact group G, the **reduced** C*-**algebra** $C_r^* G$ is the norm closure of $L^1(G)$ in the left regular representation on $L^2(G)$ (that is, when $L^1(G)$ acts by left convolutors on $L^2(G)$). The **full**, or **maximal** C*-**algebra** $C^* G$ is the completion of $L^1(G)$ characterized by the universal property that every *-representation of $L^1(G)$ on a Hilbert space, extends to $C^* G$. By the universal property, the left regular representation induces an epimorphism $\lambda_G \colon C^* G \to C_r^* G$. A famous characterization of amenability is that G is amenable if and only if λ_G is an isomorphism (see [Ped79]). Roughly speaking, we say that G is K-amenable if λ_G induces isomorphisms in K-theory:

$$(\lambda_G)_* \colon K_i(C^* G) \to K_i(C_r^* G) \qquad i = 0, 1$$

should be an isomorphism. For technical reasons, one defines G to be **K-amenable** when the unit element of the Kasparov ring $KK_G(\mathbb{C}, \mathbb{C})$ may be expressed by means of representations of G weakly contained in the left regular representation of G. This forces $(\lambda_G)_*$ to be an isomorphism. More generally, it was proved in Proposition 3.4 of [JV84] that, if the locally compact group G is K-amenable, then for any C*-algebra A on which G acts, the canonical epimorphism $\lambda_{G,A}$ from the full crossed product $A \rtimes G$ to the reduced crossed product $A \rtimes_r G$ induces isomorphisms in K-theory:

$$(\lambda_{G,A})_* \colon K_i(A \rtimes G) \xrightarrow{\cong} K_i(A \rtimes_r G).$$

An important result of J.-L. Tu ([Tu99], see also [Jul98, Sec. 6]) is that the class of locally compact groups with the Haagerup property is contained in the class of K-amenable groups. This inclusion is strict! Indeed, the semidirect products $\mathbb{R}^2 \rtimes SL_2(\mathbb{R})$ and $\mathbb{Z}^2 \rtimes SL_2(\mathbb{Z})$ are K-amenable, by [JV84, Prop. 3.3] but, as we shall see in Section 1.4 below, relative property (T) prevents them from having the Haagerup property. However, a common feature of amenable groups, groups with the Haagerup property and K-amenable groups, is the fact that their closed subgroups with property (T) are necessarily compact. In the final chapter of this book, we shall discuss some classes of K-amenable groups for which the Haagerup property is unknown.

1.3.3 The Baum–Connes conjecture

The most spectacular result about the Haagerup property was recently obtained by N. Higson and G. Kasparov [HK97] and Tu [Tu99]: the Baum–Connes conjecture holds for groups with the Haagerup property. Without going into details, we recall that the Baum–Connes conjecture is a tantalizing programme that identifies two objects associated with a locally compact group G, one geometrical or topological, and one analytical. The topological side is the G-equivariant K-homology with G-compact supports $K_i^G(\underline{EG})$ of the classifying space $\underline{EG}$ for G-proper actions. The analytical side is the K-theory $K_i(C_r^*G)$ of the reduced C*-algebra C_r^*G, which already appeared in 1.3.2. The **Baum–Connes conjecture** for G [BCH94] is the statement that the analytic assembly map, or index map,

$$\mu_i^G \colon K_i^G(\underline{EG}) \to K_i(C_r^*G)$$

is an isomorphism when $i = 0$ or 1. For groups with the Haagerup property, Higson and Kasparov [HK97] and Tu [Tu99] actually prove a more general version of the conjecture, where one takes coefficients in an auxiliary G-C*-algebra, and which computes $K_i(A \rtimes_r G)$ (see also [Jul98] for a survey).

One interest of the Baum–Connes conjecture is that, when restricted to the class of discrete groups, it implies several other famous conjectures in either functional analysis or topology. One consequence in functional analysis is **the conjecture of idempotents**, or **Kaplansky–Kadison conjecture** (see [Val89]): for a torsion-free group Γ with the Haagerup property, the reduced C*-algebra $C_r^*\Gamma$ has no idempotents other than 0 or 1; this follows from the surjectivity of μ_0^Γ. Since $C_r^*\Gamma$ is a completion of the complex group algebra $\mathbb{C}\Gamma$, this implies in particular that, for the same class of groups, $\mathbb{C}\Gamma$ has no nontrivial idempotents, a fact for which there is no purely algebraic proof so far (see [Pas77] for algebraic approaches to the conjecture of idempotents).

We now turn to applications of the Baum–Connes conjecture in topology. For Γ a discrete group, let $B\Gamma$ be the classifying space of Γ (so that the universal cover $E\Gamma = \widetilde{B\Gamma}$ classifies proper, free Γ-actions). Then $K_i(B\Gamma) \simeq K_i^{\Gamma}(E\Gamma)$ canonically. Since any proper, free Γ-action is in particular proper, there is a canonical Γ-equivariant "forgetful map" $\iota\colon E\Gamma \mapsto \underline{E\Gamma}$, inducing $\iota_*\colon K_i(B\Gamma) \to K_i^{\Gamma}(\underline{E\Gamma})$; the map ι_* is rationally injective (see [BCH94]). The Baum–Connes conjecture for Γ clearly implies rational injectivity of $\mu_i^{\Gamma} \circ \iota_*$; in turn, this implies the **Novikov conjecture** for Γ: higher signatures coming from $H^*(\Gamma, \mathbb{Q})$ are oriented homotopy invariants of closed oriented manifolds with fundamental group Γ.

In connection with the Novikov conjecture, a very weak form of amenability was introduced by G.-L.Yu [Yu00].

Definition 1.3.1. A finitely generated group Γ has **property A** if, for some finite generating subset S, with associated word length function $|\cdot|_S$, and for all $r, \epsilon > 0$, there exists a family $(A_\gamma)_{\gamma \in \Gamma}$ of finite subsets of $\Gamma \times \mathbb{N}$ such that

(1) $(\gamma, 1) \in A_\gamma$ for all $\gamma \in \Gamma$;
(2) $\operatorname{card}(A_\gamma \triangle A_{\gamma'}) < \epsilon \operatorname{card}(A_\gamma \cap A_{\gamma'})$, whenever $\left|\gamma^{-1}\gamma'\right|_S \leq r$;
(3) there exists $R > 0$ such that, if (x, m) and (y, n) are in the same A_γ, then $\left|x^{-1}y\right|_S \leq R$.

It is not difficult to see that, if property A holds with respect to one finite generating subset S of Γ, then it holds with respect to every such set. Notice that finitely generated amenable groups satisfy property A. Indeed, in this case we may find a finite Følner set F for the ball of radius $r > 0$ in Γ:

$$\frac{\operatorname{card}(F \triangle gF)}{\operatorname{card}(F)} < \frac{\epsilon}{2}$$

if $|g|_S \leq r$; then the sets $A_\gamma = \gamma F \times \{1\}$ do the job.

Definition 1.3.2. A finitely generated group Γ **admits a uniform embedding in Hilbert space** if, for some finite generating subset S, there exist a Hilbert space $\mathcal{H}$, a map $\beta\colon \Gamma \to \mathcal{H}$, and nondecreasing functions $\rho_1, \rho_2\colon \mathbb{R}^+ \to \mathbb{R}^+$ such that

(1) $\lim_{r \to \infty} \rho_i(r) = \infty$ when $i = 1, 2$;
(2) $\rho_1\left(\left|x^{-1}y\right|_S\right) \leq \|\beta(x) - \beta(y)\| \leq \rho_2\left(\left|x^{-1}y\right|_S\right)$ for all $x, y \in \Gamma$.

Obviously this property does not depend on the choice of the finite generating set S. It was shown by Yu ([Yu00, Thm 2.2]) that any group with property A admits a uniform embedding in Hilbert space. The same is true for a finitely generated group Γ with the Haagerup property. Indeed, if α is a proper isometric action of Γ on a Hilbert space $\mathcal{H}$, then we may take $\beta(x) = \alpha(x)(v)$, where $v \in \mathcal{H}$. In [Yu00, Cor. 1.2], it is proved that, if a finitely generated group Γ admits a uniform embedding in Hilbert space, and if $B\Gamma$ is a finite complex, then $\mu_i^\Gamma \circ \iota_*$ is rationally injective, so that the Novikov conjecture holds for Γ.

Property A is a weak form of amenability. This was made precise by Higson and J. Roe [HR00]: a finitely generated group Γ has property A if and only if the action of Γ on its Stone–Čech compactification $\beta\Gamma$ is amenable. It is a folk conjecture that every countable group Γ acts amenably on $\beta\Gamma$; however a counter-example to that has been announced by Gromov [Gro99]. From the Higson–Roe result, it follows that, if Γ acts amenably on some compact space, and if the classifying space $B\Gamma$ is a finite complex, then $\mu_i^\Gamma \circ \iota_*$ is rationally injective. More recently, Higson [Hig00] strengthened that result: appealing to results of Tu [Tu99], he showed that, if a discrete group Γ acts amenably on some compact space, then the map μ_i^Γ is injective.

1.4   What this book is about

In Chapter 2, Paul Jolissaint presents two new characterizations of the Haagerup property in terms of actions on von Neumann algebras. As announced above, Section 2.1 contains a quick proof of the equivalence of conditions (1) to (4) in Subsection 1.1.1. Section 2.2 is devoted to the case of actions on standard Borel spaces (this may be viewed as the abelian von Neumann algebra case). It is proved that a locally compact, second countable group G has the Haagerup property if and only if there exists a measure-preserving action of G on a standard probability space (S, μ), with the following properties:

(1) the action is **strongly mixing**: for all Borel subsets A, B in S,

$$\lim_{g \to \infty} \mu(Ag \cap B) = \mu(A)\mu(B);$$

(2) S has a **nontrivial asymptotically invariant sequence**: there exists a sequence of Borel sets $(A_n)_{n \geq 1}$ such that $\mu(A_n) = 1/2$ for all n and

$$\lim_{n \to \infty} \sup_{g \in K} \mu(A_n g \,\triangle\, A_n) = 0$$

for all compact subsets K of G.

Perhaps the simplest noncommutative analogue of a G-space is the hyperfinite type II_1 factor R endowed with an action of G by *-automorphisms. Thus, in Section 2.3, an analogous result is proved in the case of R, namely, G has the Haagerup property if and only if it admits an action α on R such that

(1) α is strongly mixing: for all $x, y \in R$,

$$\lim_{g \to \infty} \tau(\alpha_g(x)y) = \tau(x)\tau(y);$$

(2) there exists a sequence of projections $(e_n)_{n \geq 1} \subset R$ such that $\tau(e_n) = 1/2$ for all n and

$$\lim_{n \to \infty} \sup_{g \in K} \|\alpha_g(e_n) - e_n\|_2 = 0,$$

for every compact subset K of G.

In Chapter 3, Pierre Julg gives a new, geometric proof of the Haagerup property for $SO(n, 1)$ and $SU(m, 1)$.

In Chapter 4, Pierre-Alain Cherix, Michael Cowling and Alain Valette obtain a classification of connected Lie groups with the Haagerup property. Before stating the result precisely, we need one more definition. (We refer to [BR95, Thm 1.1] for the equivalences of the conditions.)

Definition 1.4.1. Let G be a (compactly generated) locally compact group, and let H be a closed subgroup. We say that the pair (G, H) has **relative property (T)** if the following equivalent conditions are satisfied:

(1) every continuous, conditionally negative definite function on G is bounded on H;

(2) whenever a sequence of continuous, normalized, positive definite functions on G converges to 1 uniformly on compact subsets of G, then it converges to 1 uniformly on H;

(3) if a representation of G contains the trivial representation weakly, then it has nonzero H-fixed vectors;

(4) every continuous, isometric action of G on an affine Hilbert space has a fixed point under H.

The semidirect products $(\mathbb{R}^2 \rtimes SL_2(\mathbb{R}), \mathbb{R}^2)$ and $(\mathbb{Z}^2 \rtimes SL_2(\mathbb{Z}), \mathbb{Z}^2)$ are nontrivial examples of groups with relative property (T) (see [HV89, pp. 18 and 94]).

Clearly, relative property (T) is an obstruction to the Haagerup property; more precisely, if the group G contains a *noncompact* subgroup H such that the pair (G, H) has relative property (T), then G cannot have the Haagerup property. This is the only known obstruction to the Haagerup property! We prove in Theorem 4.0.1 that, for a connected Lie group G, this really is the only one. More precisely, we show that the following are equivalent:

(1) G has the Haagerup property;

(2) if, for some closed subgroup H, the pair (G, H) has relative property (T), then H is compact;

(3) G is locally isomorphic to a direct product

$$M \times \mathrm{SO}(n_1, 1) \times \cdots \times \mathrm{SO}(n_k, 1) \times \mathrm{SU}(m_1, 1) \times \cdots \times \mathrm{SU}(m_l, 1),$$

where M is an amenable Lie group, that is, a compact extension of a solvable Lie group.

This means in particular that, as far as noncompact simple Lie groups are concerned, we have a clear-cut situation: such a group has the Haagerup property if and only if it is locally isomorphic to either $\mathrm{SO}(n, 1)$ or $\mathrm{SU}(m, 1)$; otherwise it has property (T) (by the results in [HV89]).

If a locally compact group G is endowed with a left-invariant metric, we say that a function on G is **radial** if it depends only on the distance to the origin in G. We say that G has the **radial Haagerup property** if there exists a sequence of normalized radial positive definite functions ϕ_n on G, vanishing at ∞, which converge to 1 locally uniformly. Let N be a simply connected, two-step nilpotent Lie group of Heisenberg type. Such a group admits a one-parameter group A of automorphic homogeneous dilations, so we may form the semidirect product NA, a simply connected solvable group with a natural left-invariant Riemannian metric. In Chapter 5, Michael Cowling shows that although NA is amenable, in general it does not have the radial Haagerup property. More precisely, NA has the radial Haagerup property if and only if the centre of N is one-dimensional. He uses this result to give a new proof that $\mathrm{Sp}(n, 1)$ has property (T) when $n \geq 2$.

Chapter 6, by Paul Jolissaint, Pierre Julg and Alain Valette, is mainly devoted to discrete groups. However, the chapter begins with a section devoted to general constructions preserving the class of groups with the Haagerup property. For example, this class is stable under inductive limits (which allows us to prove that some adèle groups have the Haagerup property). We also reprove a criterion from [Jol00] allowing one to lift the Haagerup property from a subgroup to the ambient group. In Section 6.2, we consider discrete groups acting on trees: we prove that if a group acts on a tree in such a way that vertex-stabilizers have the Haagerup property, and edge-stabilizers are finite, then the group has the Haagerup property. In particular, the class of discrete groups with the Haagerup property is stable under free products and amalgamated products over finite groups, a fact already proved in [Jol00]. Finally, in Section 6.3, we deal with finitely presented groups, and establish a criterion for a group presentation which implies that the corresponding group has the Haagerup property and simultaneously admits a classifying space which is a finite two-dimensional complex.

Chapter 7, by Alain Valette, gives a list of open questions and ties up some loose ends. In particular, we discuss the behaviour of the Haagerup property under central extensions.

We thank E. Ghys, A. Nevo, H. Oh, Y. Shalom, K. Taylor and especially B. Bekka for a number of useful conversations and comments.

Chapter 2

Dynamical Characterizations
by Paul Jolissaint

2.1 Definitions and statements of results

Before discussing the relationship between the Haagerup property and group actions on von Neumann algebras, we present a proof of the equivalence of the four characterizations of this property stated in Chapter 1. The equivalences are spread over [AW81], [BCV95], [Jol00] and [Jul98], and it may be useful to gather them all together in the same place.

Theorem 2.1.1. *For a locally compact, second countable noncompact group G, the following conditions are equivalent:*

(1) *[AW81] there exists a proper, continuous function $\psi\colon G \to \mathbb{R}^+$ which is conditionally negative definite, that is, $\psi(g^{-1}) = \psi(g)$ for all $g \in G$, and for all $g_1, \dots, g_n \in G$ and all $a_1, \dots, a_n \in \mathbb{C}$ with $\sum a_i = 0$,*

$$\sum_{i,j} \overline{a}_i a_j \psi(g_i^{-1} g_j) \leq 0;$$

(2) *the abelian C^*-algebra $C_0(G)$ possesses an approximate unit of normalized, positive definite functions, that is, there exists a sequence $(\varphi_n)_{n \geq 1}$ of functions in $C_0(G)$ such that $\varphi_n(e) = 1$ for all n, $\varphi_n \to 1$ uniformly on compact subsets of G and which are positive definite, that is,*

$$\sum_{i,j} \overline{a}_i a_j \varphi_n(g_i^{-1} g_j) \geq 0$$

for all $g_1, \dots, g_n \in G$ and all $a_1, \dots, a_n \in \mathbb{C}$;

© Springer Basel 2001
P.-A. Cherix et al., *Groups with the Haagerup Property*,
Modern Birkhäuser Classics, DOI 10.1007/978-3-0348-0906-1_2

(3) *[Jol00] there exists a C_0 unitary representation $(\pi, \mathcal{H})$ of G, that is, all matrix coefficients $\varphi_{\xi, \eta} \colon g \mapsto \langle \pi(g)\xi, \eta \rangle$ belong to $C_0(G)$, which weakly contains the trivial representation 1_G (denoted $1_G \prec \pi$);*

(4) *[BCV95] G is **a-T-menable**: there exists a Hilbert space $\mathcal{H}$ and an isometric affine action α of G on $\mathcal{H}$ which is proper in the sense that, for all pairs of bounded subsets B and C of $\mathcal{H}$, the set of elements $g \in G$ such that $\alpha_g(B) \cap C \neq \emptyset$ is relatively compact.*

Moreover, if these conditions hold, one can choose in (1) a proper, continuous, conditionally negative definite function ψ such that $\psi(g) > 0$ for all $g \neq e$, and similarly the representation π in condition (3) may be chosen such that for all $g \neq e$, there exists a unit vector $\xi \in \mathcal{H}$ such that $|\langle \pi(g)\xi, \xi \rangle| < 1$. In particular, π is faithful.

Proof. First we prove the equivalence of (1) and (2). If ψ satisfies (1), Schoenberg's theorem ([HV89, p. 66]) states that $\exp(-t\psi)$ is positive definite for all positive t. Hence (1) implies (2).

Conversely, if G satisfies (2), let $(K_n)_{n \geq 1}$ be an increasing sequence of compact subsets of G whose union is G. Choose an unbounded increasing positive sequence $(\alpha_n)_{n \geq 1}$ and a decreasing sequence $(\varepsilon_n)_{n \geq 1}$, tending to 0, such that $\sum \alpha_n \varepsilon_n$ converges. For all n, choose a continuous, positive definite function φ_n on G such that $\varphi_n \in C_0(G)$ and

$$\sup_{g \in K_n} |\varphi_n(g) - 1| \leq \varepsilon_n.$$

Replacing φ_n by $|\varphi_n|^2$ if necessary, we assume further that $0 \leq \varphi_n \leq 1$ for all n. Set, for $g \in G$,

$$\psi(g) = \sum_{n \geq 1} \alpha_n (1 - \varphi_n(g)),$$

which defines a conditionally negative definite function on G. As the series converges uniformly on compact sets, ψ is continuous. To check that it is proper, take $R > 0$, and fix an integer n so large that $\alpha_n \geq 2R$. As φ_n belongs to $C_0(G)$, there exists a compact subset L of G such that $|\varphi_n(g)| < 1/2$ for all $g \notin L$. Then

$$\{g \in G : \psi(g) \leq R\} \subseteq \{g \in G : 1 - \varphi_n(g) \leq 1/2\} \subseteq L.$$

Now we prove the equivalence of (1) and (4). Let ψ be a (not necessarily proper) continuous, conditionally negative definite function on G. Then, by [HV89, p. 63], there exists an essentially unique triple $(\mathcal{H}, \pi, b)$ where $\mathcal{H}$ is a real Hilbert space, π is an orthogonal representation of G on $\mathcal{H}$ and b is a

π-cocycle (that is, $b(gh) = b(g) + \pi(g)b(h)$ for all $g, h \in G$), $\mathcal{H}$ is topologically generated by the range of b, and finally

$$\psi(g) = \|b(g)\|^2$$

for all g. The associated affine action is defined by

$$\alpha_g(\xi) = \pi(g)\xi + b(g).$$

Conversely, if b is a cocycle as above, the function $g \mapsto \|b(g)\|^2$ defines a continuous, conditionally negative definite function. The equivalence of (1) and (4) follows from the fact that ψ is proper if and only if α is.

Now we show that (2) and (3) are equivalent. If $(\varphi_n)_{n \geq 1}$ satisfies condition (2), let $(\pi_n, \mathcal{H}_n, \xi_n)$ be the Gel'fand–Naimark–Segal triple associated with φ_n, and set

$$\pi = \bigoplus_n \pi_n.$$

Then π is a C_0-representation and $1_G \prec \pi$. Conversely, if $(\pi, \mathcal{H})$ satisfies condition (3), let $(\eta_n)_{n \geq 1}$ be a sequence of unit vectors such that, for all compact subsets K of G,

$$\lim_{n \to \infty} \sup_{g \in K} \|\pi(g)\eta_n - \eta_n\| = 0.$$

Set $\varphi_n = \langle \pi(\cdot)\eta_n, \eta_n \rangle$. Then the sequence (φ_n) satisfies condition (2).

In order to prove the additional property of ψ in condition (1), choose a sequence $(V_n)_{n \geq 1}$ of relatively compact neighbourhoods of the identity e such $\bigcap_n V_n = \{e\}$. Next, choose nonnegative continuous functions f_n for all n such that $\int_G f_n(g)^2 dg = 1$ and such that the associated matrix coefficient function

$$\omega_n(g) = \langle \lambda_G(g)f_n, f_n \rangle$$

is supported in V_n. Replacing ψ by

$$\psi' = \psi + \sum_{n \geq 1} \frac{1}{2^n}(1 - \omega_n),$$

it is easy to check that $\psi'(g) > 0$ for all $g \neq e$.

A similar argument works for the representation π in condition (3).   □

Remark. In Definition 2.4 of [BR88], V. Bergelson and J. Rosenblatt proved that if G satisfies condition (2) of Theorem 2.1.1, then it has the following interesting property: fix an infinite dimensional, separable Hilbert space $\mathcal{H}$, and denote by $\mathrm{Rep}(\mathcal{H})$ the set of all unitary representations of G on $\mathcal{H}$, endowed with a suitable natural topology. Then the subset of C_0-representations is dense in $\mathrm{Rep}(\mathcal{H})$.

In the 1980s, Rosenblatt [Ros81], K. Schmidt [Sch81] and A. Connes and B. Weiss [CW80] found characterizations of amenability and of property (T) for countable groups in terms of measure preserving ergodic actions. For instance, it follows from [Sch81] and [CW80] that a countable group Γ is amenable if and only if no measure preserving ergodic action of Γ is strongly ergodic, and on the other hand, Γ has property (T) if and only if every measure preserving ergodic action of Γ is strongly ergodic (this means that there are no nontrivial asymptotically invariant sequences: see the remark following the proof of Theorem 2.2.2 and Proposition 2.2.3). The main results of this chapter fit into this circle of ideas, since they characterize the Haagerup property in terms of suitable measure preserving ergodic actions on the one hand, and on some approximately finite dimensional factors on the second. In order to state them, we need to fix notation and give some definitions.

Assume that G is a locally countable second countable group that acts (on the right) on a standard probability space (S, μ) by measure-preserving Borel automorphisms.

Definition 2.1.2. The action of G on (S, μ) is said to be **strongly mixing** if, for all Borel subsets A and B of S,

$$\lim_{g \to \infty} \mu(Ag \cap B) = \mu(A)\mu(B),$$

that is, for all positive ε, there exists a compact subset K of G such that

$$|\mu(Ag \cap B) - \mu(A)\mu(B)| < \varepsilon \qquad \forall g \in G \setminus K.$$

A sequence of Borel subsets $(A_n)_{n \geq 1}$ of S is said to be **asymptotically invariant** if, for all compact subsets K of G,

$$\sup_{g \in K} \mu(A_n g \triangle A_n) \to 0 \quad \text{as } n \to \infty.$$

It is said to be **nontrivial** if moreover

$$\inf_n \mu(A_n)(1 - \mu(A_n)) > 0.$$

A sequence of nonnull Borel subsets $(A_n)_{n \geq 1}$ of S is said to be a **Følner sequence** if $\mu(A_n) \to 0$ as $n \to \infty$ and if for all compact subsets K of G,

$$\sup_{g \in K} \frac{\mu(A_n g \triangle A_n)}{\mu(A_n)} \to 0 \quad \text{as } n \to \infty.$$

Finally, we define the unitary representation $\pi_S \colon G \to \mathcal{U}(L^2(S, \mu))$ by

$$(\pi_S(g)\xi)(s) = \xi(sg),$$

for all $\xi \in L^2(S,\mu)$, $g \in G$ and $s \in S$, and the subspace $L_0^2(S,\mu)$ to be

$$\left\{ \xi \in L^2(S,\mu) : \int_S \xi \, d\mu = 0 \right\}.$$

Then $L_0^2(S,\mu)$ is closed and G-invariant, and we denote by ρ_S the restriction of π_S to $L_0^2(S,\mu)$.

Remark. Similar sequences have already been used several times: see [CW80], [Rin88], [Ros81], [Sch81] and [Sch80]. For instance, a Følner sequence in our sense is called an **I-sequence** in [Sch81].

There are relationships between the existence of nontrivial asymptotically invariant sequences and the existence of Følner sequences; for simplicity, assume that G is countable. If (S,μ) is an ergodic G-space which has a nontrivial asymptotically invariant sequence, then for all $c \in \,]0,1[$, there exists an asymptotically invariant sequence $(C_n)_{n \geq 1}$ such that $\mu(C_n) = c$ for all n (see [Sch81]), and this obviously implies the existence of Følner sequences. However, the converse fails: Example 2.7 of [Sch81] exhibits an action of the nonabelian free group F_3 on a probability space (S,μ) with no nontrivial asymptotically invariant sequences, but with a Følner sequence.

Here is our first main result.

Theorem 2.1.3. *Let G be a locally compact second countable group. Then G has the Haagerup property if and only if there exists a measure preserving G-action on a standard probability space (S,μ) such that*

(1) *the action of G on (S,μ) is strongly mixing, and*
(2) *(S,μ) contains a Følner sequence for the G-action.*

Moreover, (S,μ) contains a nontrivial asymptotically invariant sequence, and S may be taken to be a compact metrizable space with a continuous, essentially free action of G.

This will be proved in Section 2.2, where the construction of (S,μ) is taken from [Sch96].

Remark. We will see in the next section that conditions (1) and (2) translate into properties of the representation ρ_S, namely, condition (1) holds if and only if ρ_S is of class C_0, and condition (2) is equivalent to the condition that $1_G \prec \rho_S$ by [Rin88, Prop. 4]. This means that the representation π in condition (3) of Theorem 2.1.1 may be chosen to be of the form ρ_S.

It turns out that Theorem 2.1.3 has a noncommutative analogue. In order to state it, we need more notation and definitions. Let N be a von Neumann algebra with separable predual, and let φ be a faithful normal state on N. We denote by $\|x\|_\varphi$ the associated Hilbert norm $\varphi(x^*x)^{1/2}$. By completing N with respect to $\|\cdot\|_\varphi$ and extending the left multiplication on N, we obtain a Hilbert space $L^2(N, \varphi)$ on which N acts. Moreover, we assume that there is an action $\alpha\colon G \to \mathrm{Aut}(N)$ such that φ is α-invariant: $\varphi \circ \alpha_g = \varphi$ for all $g \in G$.

The following is the noncommutative analogue of Definition 2.1.2.

Definition 2.1.4. With the same assumptions as above, we say that the action α is **strongly mixing** for φ if

$$\lim_{g\to\infty} \varphi(\alpha_g(x)y) = \varphi(x)\varphi(y)$$

for all $x, y \in N$. A sequence of projections $(e_k)_{k\geq 1}$ on N is said to be a **nontrivial asymptotically invariant sequence** for α and φ if

$$\lim_{k\to\infty} \sup_{g\in K} \|\alpha_g(e_k) - e_k\|_\varphi = 0$$

for all compact subsets K of G and

$$\lim_{k\to\infty} \varphi(e_k)(1 - \varphi(e_k)) > 0.$$

The sequence of projections is said to be a **Følner sequence** if

$$\lim_{k\to\infty} \sup_{g\in K} \frac{\|\alpha_g(e_k) - e_k\|_\varphi}{\|e_k\|_\varphi} = 0$$

for all compact subsets K of G and

$$\lim_{k\to\infty} \varphi(e_k) = 0.$$

In Section 2.3, we observe first that if α is strongly mixing for φ and if N contains a Følner sequence, then G has the Haagerup property. Our second main result is a converse.

Theorem 2.1.5. *Let G be a locally compact second countable group with the Haagerup property. Then for each factor N listed below, there exist an action α of G on N and an α-invariant state φ for which α is strongly mixing, and N contains a Følner sequence and a nontrivial asymptotically invariant sequence for α and φ:*

(1) *N is the hyperfinite factor R of type II_1 and φ is the canonical trace τ.*

(2) *N is the approximately finite dimensional factor $R_{0,1} = R \otimes B$ of type II_∞ and $\varphi = \tau \otimes \omega$, where B is the type I_∞ factor and ω is a suitable normal state on B.*

(3) *N is the Powers factor R_λ of type III_λ and $\varphi = \varphi_\lambda$ is the associated Powers state.*

This will be proved in Section 2.3. The idea of the proof of Theorem 2.1.5 is as follows: given a C_0-representation $(\pi, \mathcal{H})$ of G such that $1_G \prec \pi$, one defines an action of G on the canonical anticommutation relation C*-algebra CAR$(\mathcal{H})$, and every factor listed above is obtained from the Gel'fand–Naimark–Segal construction starting from a suitable state on CAR$(\mathcal{H})$.

Remark. As in the commutative case, the existence and the mixing properties of the asymptotically invariant sequences and the Følner sequences in Theorem 2.1.5 may be expressed in terms of a representation ρ of G on a suitable subspace of $L^2(N)$. See Section 2.3 for a precise statement.

Finally, we focus on actions of a *discrete* group G with the Haagerup property on the hyperfinite II$_1$-factor R and improve Theorem 2.1.5: there exists an action α of G on R which has many nontrivial asymptotically invariant sequences. See the end of Section 2.3, where we discuss centralizing sequences, the asymptotic centralizer R_ω, where ω is a free ultrafilter on $\mathbb{N}$, and the induced action α^ω on R_ω (see also [Con75] and [Ocn85]).

Theorem 2.1.6. *Let G be a countable group with the Haagerup property. Then there exists an action α of G on R with the following properties:*

(1) *α is strongly mixing and (centrally) free;*
(2) *the fixed point algebra $(R_\omega)^\alpha$, that is, the set of all $x \in R_\omega$ such that $\alpha_g^\omega(x) = x$ for all $g \in G$, is a type II$_1$ subalgebra of R_ω.*

This result merits some comment. In [Ocn85], A. Ocneanu classified (centrally) free actions of countable *amenable* groups on R, and to do that, he used the fact that the fixed point algebra $(R_\omega)^\alpha$ is always of type II$_1$ for such an action. On the other hand, for any countable *nonamenable* group G, V.F.R. Jones [Jon83] studied the Bernoulli shift action β of G on the space R, realized as the infinite tensor product

$$R = \bigotimes_{g \in G} M_2(\mathbb{C}).$$

It is obvious that β is strongly mixing, and Jones proved that the fixed point algebra $(R_\omega)^\beta$ is trivial. Theorem 2.1.6 shows that groups with the Haagerup property behave in some sense like amenable groups.

2.2 Actions on measure spaces

The following result is a special case of Proposition 2.3.1 in the next section.

Proposition 2.2.1. *Assume that the locally compact second countable group G has a measure-preserving action on a standard probability space (S, μ), that the* *action is strongly mixing and that there exists a Følner sequence in S. Then G has the Haagerup property.*

We simply observe that the strong mixing property is equivalent to the fact that the representation ρ_S is a C_0-representation and that the existence of a Følner sequence implies $1_G \prec \rho_S$.

The main result of this section is the following converse.

Theorem 2.2.2. *Suppose that G has the Haagerup property. Then there exists a measure-preserving Borel action of G on a standard probability space (S, μ), with the following properties:*

(1) *the action is strongly mixing;*

(2) *there exists a sequence of Borel sets $(A_n)_{n \geq 1}$ in S such that $\mu(A_n) = 1/2$ for all n and for all compact subsets K of G,*

$$\lim_{n \to \infty} \sup_{g \in K} \mu(A_n g \triangle A_n) = 0;$$

(3) *the action is essentially free: the subset of all $s \in S$ for which the stabilizer $G_s \neq \{e\}$ is of measure zero.*

Before proving Theorem 2.2.2, we introduce more notation. Let $(\pi, \mathcal{H})$ be a representation of G, and set $\mathcal{H}^\sigma = \sum_{k \geq 0}^\oplus \mathcal{H}^{\times k}$, where $\mathcal{H}^{\times 0} = \mathbb{C}$, and for $k > 0$, $\mathcal{H}^{\times k}$ is the kth symmetric tensor product of $\mathcal{H}$, that is, the closed subspace of the Hilbert tensor product space $\mathcal{H}^{\otimes k}$ generated by the vectors of the form

$$\sum_{\varepsilon \in S_k} \xi_{\varepsilon(1)} \otimes \dots \otimes \xi_{\varepsilon(k)},$$

where S_k denotes the usual permutation group. Then the representation π extends in a natural way to a representation π^σ of G on $\mathcal{H}^\sigma$ which leaves the subspace $\mathcal{H}_0^\sigma = \mathcal{H}^\sigma \ominus \mathcal{H}^{\times 0}$ invariant. Finally, we denote by π_0^σ the restriction of π^σ to $\mathcal{H}_0^\sigma$.

Proof of Theorem 2.2.2. Fix a proper, continuous, conditionally negative definite function $\psi \colon G \to [0, +\infty[$, nonvanishing except at e. For $n \geq 1$, set $\varphi_n = \exp(-\psi/n)$, and denote by $(\pi_n, \mathcal{H}_n, \xi_n)$ the associated Gel'fand–Naimark–Segal triple. Since φ_n is real-valued, there is a real Hilbert subspace $\mathcal{H}_n'$ of $\mathcal{H}_n$, containing ξ_n, such that

$$\mathcal{H}_n = \mathcal{H}_n' \oplus i\mathcal{H}_n' \quad \text{and} \quad \pi_n(g)\mathcal{H}_n' = \mathcal{H}_n'$$

for all n and g. Write $\mathcal{H}$ for $\bigoplus_{n \geq 1} \mathcal{H}_n$ and π for $\bigoplus_{n \geq 1} \pi_n$, and observe that $\mathcal{H} = \mathcal{H}' \oplus i\mathcal{H}'$, where $\mathcal{H}' = \bigoplus_{n \geq 1} \mathcal{H}_n'$. To simplify notation, we denote by ξ_n

the corresponding vector $0 \oplus \ldots \oplus \xi_n \oplus \ldots \in \mathcal{H}'$, and we observe that $\xi_n \perp \xi_m$ when $n \neq m$.

We first define a G-space (Ω, ν), as in [Sch96, Sec. 3], satisfying conditions (1) and (2), and then we indicate how to obtain a G-space (S, μ) which also satisfies condition (3), using the techniques of S. Adams, G.A. Elliot and T. Giordano [AEG94, Prop. 1.2] (faithfulness of π is required here).

Choose a countable orthonormal basis $\mathcal{B}$ of $\mathcal{H}'$ containing $\{\xi_n : n \geq 1\}$, and set

$$(\Omega, \nu) = \prod_{b \in \mathcal{B}} \left(\mathbb{R}, \frac{1}{\sqrt{2\pi}} \exp\left(-\frac{x^2}{2}\right) dx \right).$$

Define $\eta' : \mathcal{H}' \to \Omega$ by $\eta'(\xi) = (\langle \xi, b \rangle)_{b \in \mathcal{B}}$ and $X_b : \Omega \to \mathbb{R}$ by $X_b((\omega_{b'})_{b' \in \mathcal{B}}) = \omega_b$ for every $b \in \mathcal{B}$. The random variables X_b are independent Gaussians with mean 0 and variance 1. In particular, if $b_1, \ldots b_k \in \mathcal{B}$ are distinct and $l_1, \ldots, l_k$ are nonnegative integers, then

$$\int_\Omega X_{b_1}^{l_1} \ldots X_{b_k}^{l_k} \, d\nu = \prod_{j=1}^k \frac{1}{\sqrt{2\pi}} \int_\mathbb{R} x^{l_j} \exp\left(-\frac{x^2}{2}\right) dx.$$

This allows us to define $\eta : \mathcal{H}' \to L^2(\Omega, \nu, \mathbb{R})$ by

$$\eta\left(\sum_{b \in \mathcal{B}} \xi_b b \right) = \sum_{b \in \mathcal{B}} \xi_b X_b.$$

Thus η is an isometry that extends to a unitary operator $u : \mathcal{H}^\sigma \to L^2(\Omega, \nu)$ which sends $\mathcal{H}^{\times 0}$ onto the subspace of constant functions on Ω and such that $u\left(\sum_{\varepsilon \in S_n} b_{\varepsilon(1)} \otimes \ldots \otimes b_{\varepsilon(n)} \right) = X_{b_1} \ldots X_{b_n}$ for all $b_1, \cdots, b_n \in \mathcal{B}$. Moreover there exists a ν-preserving action of G on (Ω, ν) such that $u^* \pi_\Omega(g) u = \pi^\sigma(g)$ and $u^* \rho_\Omega(g) u = \pi_0^\sigma(g)$ for all $g \in G$. It is straightforward to check that π_0^σ is a C_0-representation, which shows that condition (1) is satisfied.

As a second step, we construct a sequence (B_n) of Borel subsets of Ω satisfying condition (2); our construction is inspired by E. Glasner and B. Weiss [GW97, Thm 2]. For $n \geq 1$, set $X_n = X_{\xi_n}$ (recall that $\xi_n \in \mathcal{B}$), and $X_n^g = \pi_\Omega(g) X_n$ for all $g \in G$. Observe that

$$X_n^g = \rho_\Omega(g) X_n = u \pi_0^\sigma(g) \xi_n = u \pi(g) \xi_n = \sum_{b \in \mathcal{B}} \langle \pi(g)\xi_n, b \rangle X_b.$$

Define $B_n = \{\omega \in \Omega : X_n(\omega) \geq 0\}$; by symmetry, we have $\nu(B_n) = 1/2$ for all n. Now fix $g \neq e$. Then X_n^g is a Gaussian random variable with zero mean and variance 1 because it has the same probability distribution as X_n; indeed, for all Borel subsets E of $\mathbb{R}$,

$$\nu(\{X_n^g \in E\}) = \nu(\{X_n \in E\} g^{-1}) = \nu(\{X_n \in E\}).$$

Write $X_n^g = \cos(\alpha_n(g))X_n + \sin(\alpha_n(g))Y_n$, where $\alpha_n(g) = \arccos(\langle \pi(g)\xi_n, \xi_n \rangle)$ and

$$Y_n = \frac{1}{\sin(\alpha_n(g))} \sum_{b \neq \xi_n} \langle \pi(g)\xi_n, b \rangle X_b$$

which is a Gaussian random variable with mean 0 and variance 1, and it is independent of X_n since all linear combinations of $\{X_b : b \neq \xi_n\}$ are. As in the proof of [GW97, Thm 2],

$$\nu(B_n g \triangle B_n) = \frac{\alpha_n(g)}{\pi},$$

which tends to 0 uniformly on compact sets. Thus (Ω, ν) satisfies conditions (1) and (2), but perhaps not (3), the essential freeness of the action. But, arguing as in [AEG94], define

$$(S, \mu) = \prod_{m \geq 1} (\Omega, \nu)$$

with the action $(\omega_m)g = (\omega_m g)$. Then the action is strongly mixing and essentially free, and the sequence (A_n) defined by

$$A_n = B_n \times \prod_{m \geq 2} \Omega$$

satisfies condition (3).   □

Remark. The construction of the G-space (S, μ) in the proof of Theorem 2.2.2 provides another proof of the theorem of Connes and Weiss [CW80] which works even when G is not countable. Theorem 2′ of [GW97] also provides a proof of the theorem of Connes and Weiss.

Proposition 2.2.3 ([CW80], [GW97]). *Let G be a locally compact second countable group which does not have Kazhdan's property (T). There exists a measure-preserving G-action on a probability space (S, μ) such that*

(1) *the action is ergodic and essentially free, and*

(2) *there exists a nontrivial asymptotically invariant sequence of Borel subsets of S.*

Proof. Since G does not have property (T), by [AW81], there exists a continuous, unbounded, conditionally negative definite function $\psi \colon G \to \mathbb{R}^+$; we assume that $\psi(g) > 0$ for all $g \neq e$, and we construct (S, μ) and a sequence (A_n) of Borel subsets of S as in the proof of Theorem 2.2.2. The action is essentially free and (A_n) is a nontrivial asymptotically invariant sequence. Since ψ is unbounded, it follows (for instance, from [Jol93, Lem. 4.4]) that Godement's mean $M(e^{-t\psi}) = 0$ for all positive t. By Theorem A.1 of [GW97], this implies that the G-action on S is ergodic (and even weakly mixing).   □

2.3 Actions on factors

Let N and φ be as at the end of Section 2.1. Let $(L^2(N), J, L^2(N)^+)$ be the standard form of N, and denote by $\xi_0 \in L^2(N)^+$ the associated cyclic vector, so that $\varphi(x) = \langle x\xi_0, \xi_0 \rangle$ for all $x \in N$. An action α of G on N gives rise to a unique representation $\pi \colon G \to \mathcal{U}(L^2(N))$ such that

$$\pi(g)x\pi(g^{-1}) = \alpha_g(x) \qquad \forall g \in G \quad \forall x \in N,$$

and

$$\pi(g)J = J\pi(g) \quad \text{and} \quad \pi(g)L^2(N)^+ = L^2(N)^+ \qquad \forall g \in G.$$

Moreover, if φ is α-invariant, ξ_0 is invariant under π, and the subspace $L_0^2(N)$ of all $\xi \in L^2(N)$ which are orthogonal to ξ_0 is invariant under π, and we denote by ρ the restriction of π to $L_0^2(N)$.

Proposition 2.3.1. *Assume that $\alpha \colon G \to \mathrm{Aut}(N)$ is a strongly mixing action for φ and that N contains a Følner sequence for α and φ. Then the representation ρ weakly contains the trivial representation of G and is C_0. In particular, G has the Haagerup property.*

Proof. The set of vectors $\{x\xi_0 - \varphi(x)\xi_0 : x \in N\}$ is total in $L_0^2(N)$, and for all $x, y \in N$,

$$\langle \rho(g)(x\xi_0 - \varphi(x)\xi_0), y\xi_0 - \varphi(y)\xi_0 \rangle = \varphi(\alpha_g(x)y) - \varphi(x)\varphi(y) \to 0$$

as $g \to \infty$. This proves that ρ is of class C_0. Moreover, if $(e_k)_{k \geq 1}$ is a Følner sequence for α and φ, set

$$\xi_k = \frac{e_k - \varphi(e_k)}{\sqrt{\varphi(e_k) - \varphi(e_k)^2}} \xi_0.$$

Then $\xi_k \in L_0^2(N)$ and $\|\xi_k\|_\varphi = 1$. As $\sup \varphi(e_k) < 1$, there exists a positive constant c such that $1 - \varphi(e_k) \geq c$ for all k. Hence, if K is a compact subset of G,

$$\sup_{g \in K} \|\rho(g)\xi_k - \xi_k\|_\varphi \leq c^{-1/2} \sup_{g \in K} \frac{\|\alpha_g(e_k) - e_k\|_\varphi}{\|e_k\|_\varphi} \to 0$$

as $k \to \infty$. This proves that $1_G \prec \rho$. $\qquad\square$

We describe now our realizations of the factors, states and actions listed in Theorem 2.1.5. Let $(\pi, \mathcal{H})$ be a separable unitary representation of G; we assume that the scalar product on $\mathcal{H}$ is antilinear in the first variable. Following Chapters 7 and 8 of [HJ], let A be the Fermion C*-algebra or CAR-algebra $\mathrm{CAR}(\mathcal{H})$ over $\mathcal{H}$; it is the C*-algebra generated by $\{a(\xi) : \xi \in \mathcal{H}\}$, where $a \colon \mathcal{H} \to A$ is a linear isometry, and the following **canonical anticommutation relations** hold:

$$a^*(\xi)a(\eta) + a(\eta)a^*(\xi) = \langle \xi, \eta \rangle \tag{2.1}$$

$$a(\xi)a(\eta) + a(\eta)a(\xi) = 0 \tag{2.2}$$

for all ξ and η. It is well known that A is a uniformly hyperfinite algebra, and that the representation π induces an action α of G on A characterized by

$$\alpha_g(a(\xi)) = a(\pi(g)\xi) \qquad \forall g \in G \quad \forall \xi \in \mathcal{H}.$$

Our constructions rest on the following fact (see [HJ, Thm 8.2]): for all $b \in \mathcal{L}(\mathcal{H})$ such that $0 \le b \le 1$, there exists a unique state ϕ_b on A such that

$$\phi_b(a^*(\xi_m) \dots a^*(\xi_1)a(\eta_1) \dots a(\eta_n)) = \delta_{n,m} \det(\langle \xi_j, b\eta_k \rangle), \tag{2.3}$$

for all $\xi_1, \dots, \xi_m$ and $\eta_1, \dots, \eta_n \in \mathcal{H}$. Such a state is called **quasifree**. We denote by N_b the von Neumann algebra obtained by applying the Gel'fand–Naimark–Segal construction to the pair (A, ϕ_b); it is known that N_b is always a factor. If $b = 1/2$ then $\tau = \phi_{1/2}$ is the unique normalized trace on A and the associated factor is the hyperfinite II_1 factor R. If $t \in \,]0, 1/2[$ and $\lambda = t/(1-t)$, then N_t is the Powers factor R_λ of type III_λ and ϕ_t is the Powers state, henceforth denoted by φ_λ. Finally, if $b = 1$, then ϕ_1 is the vacuum state ω; it is pure, and thus the associated factor is the type I_∞ factor B. Whenever b belongs to the commutant $\pi(G)'$ of $\pi(G)$, ϕ_b is α-invariant and the action α extends to an action of G on N_b, still denoted α. Notice that this is the case for all the values of b above.

The following result proves that there are actions of groups with the Haagerup property on the factors R and R_λ, as stated in Theorem 2.1.5.

Theorem 2.3.2. *Let G be a noncompact, locally compact second countable group.*

(1) *If $(\pi, \mathcal{H})$ is a C_0-representation of G, then for all $b \in \pi(G)'$ such that $0 \le b \le 1$, the associated action α on N_b is strongly mixing for ϕ_b.*

(2) *Assume further that G has the Haagerup property. Then there exist sequences of projections $(e_n)_{n \ge 1}$ and $(f_n)_{n \ge 1}$ in the Fermion C^*-algebra A such that, for all $t \in \,]0, 1/2]$,*

$$\phi_t(e_n) = t \quad \text{and} \quad \phi_t(f_n) = t^n$$

for all $n \ge 1$, and for all compact subsets K of G,

$$\lim_{n \to \infty} \sup_{g \in K} \|\alpha_g(e_n) - e_n\| = 0 \quad \text{and} \quad \lim_{n \to \infty} \sup_{g \in K} \frac{\|\alpha_g(f_n) - f_n\|}{\phi_t(f_n)} = 0.$$

Proof. To prove (1), it suffices to prove that $\phi_b(\alpha_g(x)y) \to \phi_b(x)\phi_b(y)$ as $g \to \infty$ for x and y in the total subset

$$\{a^*(\xi_n)\ldots a^*(\xi_1)a(\eta_1)\ldots a(\eta_m) : \xi_i, \eta_j \in \mathcal{H}\}$$

of N_b. Let us fix $\xi_1, \ldots, \xi_n, \eta_1, \ldots, \eta_m, \zeta_1, \ldots, \zeta_k$ and $\omega_1, \ldots, \omega_l$ in $\mathcal{H}$, and set

$$x = a^*(\xi_n)\ldots a^*(\xi_1)a(\eta_1)\ldots a(\eta_m)$$
$$y = a^*(\zeta_k)\ldots a^*(\zeta_1)a(\omega_1)\ldots a(\omega_l).$$

Then $\phi_b(\alpha_g(x)y)$ is equal to

$$\phi_b\Big(a^*(\pi(g)\xi_n)\ldots a^*(\pi(g)\xi_1)a(\pi(g)\eta_1)\ldots a(\pi(g)\eta_m)$$
$$a^*(\zeta_k)\ldots a^*(\zeta_1)a(\omega_1)\ldots a(\omega_l)\Big).$$

We shift $a^*(\zeta_k)\ldots a^*(\zeta_1)$ to the left between $a^*(\pi(g)\xi_1)$ and $a(\pi(g)\eta_1)$, using the canonical anticommutation relation

$$a(\pi(g)\eta_i)a^*(\zeta_j) = \langle \zeta_j, \pi(g)\eta_i \rangle - a^*(\zeta_j)a(\pi(g)\eta_i).$$

This requires mk transpositions. Next, we shift $a(\pi(g)\eta_1)\ldots a(\pi(g)\eta_m)$ to the right of $a(\omega_l)$, using the relation

$$a(\pi(g)\eta_i)a(\omega_j) = -a(\omega_j)a(\pi(g)\eta_i).$$

This, in turn, requires ml transpositions. Hence, $\phi_b(\alpha_g(x)y) = \sigma(g) + \psi(g)$, where $\sigma(g)$ is a linear combination of mk terms of the type

$$\langle \zeta_i, \pi(g)\eta_j \rangle \, \phi_b(x(i,j,g)),$$

where

$$\|x(i,j,g)\| \leq \max(1, \|\xi_n\| \ldots \|\omega_l\|)$$

for all g and all i, j, and where $\psi(g)$ is equal to

$$\pm\phi_b\Big(a^*(\pi(g)\xi_n)\ldots a^*(\pi(g)\xi_1)a^*(\zeta_k)\ldots a^*(\zeta_1)$$
$$a(\omega_1)\ldots a(\omega_l)a(\pi(g)\eta_1)\ldots a(\pi(g)\eta_m)\Big).$$

For arbitrary values of n, m, k and l, $\sigma(g) \to 0$ as $g \to \infty$ since π is a C_0-representation. If $n + k \neq l + m$, then $n \neq m$ or $k \neq l$ and $\phi_b(x)\phi_b(y) = 0$ using formula (2.3). Moreover, $\psi(g) = 0$, and $\phi_b(\alpha_g(x)y) \to 0$ as $g \to \infty$. Finally, if $n + k = m + l$, using (2.3) again,

$$\psi(g) = \pm \det \begin{pmatrix} A & B(g) \\ C(g) & D \end{pmatrix}$$

where

$$A = (\langle \zeta_i, b\omega_j \rangle) \in M_{k \times l}(\mathbb{C})$$
$$B(g) = (\langle \zeta_i, \pi(g)b\eta_j \rangle) \in M_{k \times m}(\mathbb{C})$$
$$C(g) = (\langle \pi(g)\xi_i, b\omega_j \rangle) \in M_{n \times l}(\mathbb{C})$$
$$D = (\langle \xi_i, b\zeta_j \rangle) \in M_{n \times m}(\mathbb{C}).$$

As $g \to \infty$, $B(g) \to 0$ and $C(g) \to 0$, and

$$\lim_{g \to \infty} \phi_b(\alpha_g(x)y) = \pm \det \begin{pmatrix} A & 0 \\ 0 & D \end{pmatrix}.$$

If $n \neq m$ or $k \neq l$, then necessarily $n \neq m$ and $k \neq l$. Moreover, the rank of the matrix

$$\begin{pmatrix} A & 0 \\ 0 & D \end{pmatrix}$$

is not maximal, and

$$\lim_{g \to \infty} \phi_b(\alpha_g(x)y) = 0 = \phi_b(x)\phi_b(y).$$

If $n = m$ and $k = l$, the signs of $\psi(g)$ and $\phi_b(a^*(\pi(g)\xi_n) \ldots a(\pi(g)\eta_n))$ are the same, because $2nk$ sign changes occurred. From formula (2.3),

$$\lim_{g \to \infty} \phi_b(\alpha_g(x)y) = \det \begin{pmatrix} A & 0 \\ 0 & D \end{pmatrix} = \phi_b(x)\phi_b(y).$$

Now we prove (2). Fix $t \in \,]0, 1/2]$, and let $(K_n)_{n \geq 1}$ be an increasing sequence of compact subsets of G whose union is G, and $(\sigma, \mathcal{H}_\sigma)$ be a separable representation of G satisfying condition (3) of Theorem 2.1.1. Let $(\eta_k)_{k \geq 1}$ be a sequence of unit vectors in $\mathcal{H}_\sigma$ such that

$$\sup_{g \in K} \|\sigma(g)\eta_k - \eta_k\| \to 0$$

as $k \to \infty$, for all compact subsets K of G, set $\mathcal{H} = l^2(\mathbb{N}) \otimes \mathcal{H}_\sigma$ and $\pi = 1 \otimes \sigma$. Then π is still a C_0-representation and $1_G \prec \pi$. For positive integers k, n, set

$$\xi_k^n = \delta_n \otimes \eta_k,$$

where $(\delta_n)_{n\geq 1}$ is the natural basis of $l^2(\mathbb{N})$, so that for all compact subsets K of G,

$$\lim_{k\to\infty} \sup_{\substack{n\geq 1 \\ g\in K}} \|\pi(g)\xi_k^n - \xi_k^n\| = 0,$$

and $\xi_k^n \perp \xi_l^m$ for all k, l and all $n \neq m$. For $n \geq 1$, choose an integer $k(n)$ so large that

$$\sup_{g\in K_n} \|\pi(g)\xi_{k(n)}^j - \xi_{k(n)}^j\| \leq \frac{t^{n+1}}{n^2}$$

when $1 \leq j \leq n$. Set $e_n^j = a^*(\xi_{k(n)}^j)a(\xi_{k(n)}^j)$ and $e_n = e_n^1$, and write f_n for $\prod_{j=1}^n e_n^j$. Then, using relations (2.1) and (2.2) above, $e_n^1, \ldots, e_n^n$ are pairwise commuting projections of A, while e_n and f_n are projections in A satisfying $\phi_t(e_n) = t$ and $\phi_t(f_n) = t^n$ for all $t \in]0, 1/2]$. Moreover, for all $g \in K_n$ and $1 \leq j \leq n$,

$$\|\alpha_g(e_n^j) - e_n^j\| \leq 2\|\pi(g)\xi_{k(n)}^j - \xi_{k(n)}^j\| \leq \frac{2t^{n+1}}{n^2} \leq \frac{t^n}{n^2},$$

where $\|\alpha_g(e_n^j) - e_n^j\|$ means the operator norm on A. It follows that

$$\sup_{g\in K_n} \|\alpha_g(f_n) - f_n\| \leq n \sup_{\substack{g\in K_n \\ 1\leq j\leq n}} \|\alpha_g(f_n^j) - f_n^j\| \leq \frac{1}{n} t^n = \frac{1}{n} \phi_t(f_n),$$

and similarly for (e_n). $\qquad\square$

Now, let ω be the vacuum state on A and let B be the type I_∞ factor arising from the Gel'fand–Naimark–Segal construction on (A, ω). It follows from Theorem 2.3.2 that the action of G on B is strongly mixing for ω. Set $R_{0,1} = R \otimes B$ and $\varphi = \tau \otimes \omega$; then $R_{0,1}$ is the approximately finite dimensional factor of type II_∞ and the action of G is the tensor product action. Then part (2) of Theorem 2.1.5 follows from the following result whose proof is straightforward.

Proposition 2.3.3. *Let φ and ψ be normal states on von Neumann algebras M and N respectively. Assume that there are strongly mixing actions $\alpha: G \to$ $\mathrm{Aut}(M)$ and $\beta: G \to \mathrm{Aut}(N)$ for φ and ψ respectively. If M contains a nontrivial asymptotically invariant or Følner sequence of projections for α and φ, then the action $\alpha \otimes \beta$ on $M \otimes N$ is strongly mixing for $\varphi \otimes \psi$, and $M \otimes N$ contains a nontrivial asymptotically invariant or Følner sequence for $\alpha \otimes \beta$ and $\varphi \otimes \psi$.*

Remark. Actions of groups with property (T) on the hyperfinite II_1 factor R using the CAR-algebra were also considered by H. Araki and M. Choda in [AC83].

We assume for the rest of this section that G is countable and that it has the Haagerup property. In order to state our last result, we need to recall some definitions (see for instance [Ocn85]). Let M be a type II_1 factor with separable predual; we denote by τ its normal, normalized, faithful trace and by $\|a\|_2$ the associated Hilbertian norm $\tau(a^*a)^{1/2}$. Let ω be a free ultrafilter on $\mathbb{N}$. Then

$$I_\omega = \left\{ (a_n)_{n \geq 1} \in l^\infty(\mathbb{N}, M) : \lim_{n \to \omega} \|a_n\|_2 = 0 \right\}$$

is a closed two-sided ideal of the von Neumann algebra $l^\infty(\mathbb{N}, M)$ and the corresponding quotient algebra is denoted by M^ω.

We will write $[(a_n)] = (a_n) + I_\omega$ for the equivalence class of (a_n) in M^ω, and we recall that M embeds naturally into M^ω, where the image of $a \in M$ is the class of the constant sequence $(a, a, a, \dots)$. A sequence $(a_n) \in l^\infty(\mathbb{N}, M)$ is said to be a **central sequence** if

$$\lim_{n \to \infty} \|[a, a_n]\|_2 = 0$$

for all $a \in M$, where $[a, b] = ab - ba$; two central sequences (a_n) and (b_n) are said to be **equivalent** if

$$\lim_{n \to \infty} \|a_n - b_n\|_2 = 0.$$

The relative commutant of M in M^ω is denoted by M_ω; every element of M_ω is represented by a bounded sequence (x_n) such that $\lim_{n \to \omega} \|[x, x_n]\|_2 = 0$ for all $x \in M$. Any automorphism θ of M extends naturally to an automorphism θ^ω of M^ω, whose restriction to M_ω is an automorphism of M_ω; θ is said to be **centrally trivial** if $\|\theta(x_n) - x_n\|_2 \to 0$ as $n \to \infty$ for all central sequences (x_n). A discrete group action $\alpha \colon G \to \mathrm{Aut}(M)$ is said to be **centrally free** if α_g is not centrally trivial for every $g \neq e$. Recall also that every centrally trivial automorphism of the hyperfinite factor R is inner, so that every outer action on R is centrally free.

Theorem 2.3.4. *There exists an action α of G on the hyperfinite type II_1 factor R with the following properties:*

(1) α *is strongly mixing and outer;*

(2) *the fixed point algebra $(R_\omega)^\alpha$, that is, the set of all $x \in R_\omega$ such that $\alpha_g^\omega(x) = x$ for all $g \in G$, is of type II_1.*

Proof. Let $(\sigma, \mathcal{H}_\sigma)$ be a representation of G satisfying condition (3) of Theorem 2.1.1 with the additional property that for all $g \neq e$, there exists a unit vector $\eta \in \mathcal{H}_\sigma$ such that

$$|\langle \sigma(g)\eta, \eta \rangle| < 1.$$

Let $(\pi, \mathcal{H})$ be as in the proof of Theorem 2.3.2. We again realize R as the factor obtained from the Gel'fand–Naimark–Segal construction of $\text{CAR}(\mathcal{H})$ with respect to the normalized trace τ, and the action α is still determined by

$$\alpha_g(a(\xi)) = a(\pi(g)\xi)$$

for all $g \in G$ and $\xi \in \mathcal{H}$, so that α is strongly mixing.

We prove that α is centrally free. Fix an element $g \in G \setminus \{e\}$; then there exists a unit vector $\eta \in \mathcal{H}$ such that $|\langle \sigma(g)\eta, \eta \rangle| < 1$. For all $n \geq 1$, set $e_n = a^*(\delta_n \otimes \eta)a(\delta_n \otimes \eta)$, so that e_n is a projection with trace $1/2$. Then $[(e_n)]$ belongs to R_ω, and by formulae (2.1) and (2.2) we see that

$$\|\alpha_g(e_n) - e_n\|_2^2 = \frac{1}{2} - \frac{1}{2} |\langle \eta, \sigma(g)\eta \rangle|,$$

which is positive and independent of n.

We now prove that $(R_\omega)^\alpha$ is of type II_1. Arguing as in the proof of Theorem 2.2.1 of [Con75], it suffices to prove that $(R_\omega)^\alpha$ is noncommutative. Let $e \in F_1 \subset F_2 \subset \ldots$ be an increasing sequence of finite subsets of G whose union is G, and let $(\eta_n)_{n \geq 1}$ be a sequence of unit vectors in $\mathcal{H}_\sigma$ such that

$$\max_{g \in F_n} \|\sigma(g)\eta_n - \eta_n\| \leq \frac{1}{n}.$$

Set $\xi_n = \delta_n \otimes \eta_n$ and $\zeta_n = 2^{-1/2}(\delta_n + \delta_{n+1}) \otimes \eta_n$, so that

$$\max_{g \in F_n} \|\pi(g)\xi_n - \xi_n\| = \max_{g \in F_n} \|\pi(g)\zeta_n - \zeta_n\| \leq \frac{1}{n}.$$

Define $e_n = a^*(\xi_n)a(\xi_n)$ and $f_n = a^*(\zeta_n)a(\zeta_n)$, so that both $e = [(e_n)_n]$ and $f = [(f_n)_n]$ are projections of $(R_\omega)^\alpha$ of traces $1/2$. Moreover, $ef \neq fe$ because

$$\|[e_n, f_n]\|_2^2 = \frac{1}{8}$$

for all n. □

Remark. If a group G admits an action α on a type II_1 factor M with properties (1) and (2) of Theorem 2.3.4, then M is a McDuff factor (that is, M_ω is nonabelian), by [Con75, Thm 2.2.1], since M_ω is automatically of type II_1, and of course G has the Haagerup property.

Further, it is easy to construct an action of such a group G on some nonhyperfinite type II_1 McDuff factor: let N be a type II_1 nonhyperfinite factor with separable predual. Set $N_g = N$ for all $g \in G$ and

$$M = \left(\bigotimes_{g \in G} N_g \right) \otimes R,$$

which is a nonhyperfinite type II_1 McDuff factor. The action α of G is defined by $\alpha = \beta \otimes \gamma$, where β is the Bernoulli shift action as in [Jon83] and γ is the action on R given by Theorem 2.3.4. Then it is straightforward to check that α is strongly mixing and that $(M_\omega)^\alpha$ is of type II_1.

Chapter 3

Simple Lie Groups of Rank One
by Pierre Julg

Let X be a rank one Riemannian symmetric space of the noncompact type, and G be the group of isometries of X. There are four cases:

(1) X is real n-hyperbolic space ($n \geq 2$) and $G = SO_0(n, 1)$;
(2) X is complex n-hyperbolic space ($n \geq 2$) and $G = SU(n, 1)$;
(3) X is quaternionic n-hyperbolic space ($n \geq 2$) and $G = Sp(n, 1)$;
(4) X is the hyperbolic plane over the Cayley numbers, and $G = F_{4(-20)}$.

To simplify the exposition, we shall neglect the exceptional case, and consider simultaneously the three classical series. Let K be one of the three fields $\mathbb{R}$, $\mathbb{C}$ or $\mathbb{H}$, and denote $\dim_{\mathbb{R}} K$ by k. Let us equip the right vector space K^{n+1} with the hermitian form

$$\langle z, w \rangle = \overline{z}_0 w_0 - \sum_{i=1}^{n} \overline{z}_i w_i.$$

Then the symmetric space X is defined as the open subset of the projective space $P^n(K)$ defined by the inequality $\langle x, x \rangle > 0$, and the group G is the connected component of the identity in the group of right K-linear transformations of K^{n+1} preserving the above hermitian form.

The kn-dimensional manifold X is equipped with a G-invariant Riemannian structure, and the geodesic distance between $z, w \in X$ is given by the formula

$$\cosh d(z, w) = \frac{|\langle z, w \rangle|}{\langle z, z \rangle^{1/2} \langle w, w \rangle^{1/2}}.$$

The visual boundary of X, denoted ∂X, is the subset of $P^n(K)$ defined by the equality $\langle x, x \rangle = 0$. It is a sphere of dimension $kn - 1$, equipped with

© Springer Basel 2001
P.-A. Cherix et al., *Groups with the Haagerup Property*,
Modern Birkhäuser Classics, DOI 10.1007/978-3-0348-0906-1_3

its smooth structure. For any $x \in X$, let μ_x be the visual measure on ∂X seen from x, that is, the volume form of mass 1 obtained from the canonical volume form on the unit sphere at x by the visual map. In other words, for any choice of $x \in X$, there is an identification of ∂X with a sphere S^{kn-1}, whose canonical normalized volume form gives rise to the volume form μ_x on ∂X.

3.1 The Busemann cocycle and the Gromov scalar product

The boundary ∂X carries a rich geometric structure.

Proposition 3.1.1. *Let x and y be two points in X. The difference of the distances to x and y, that is, the function*

$$z \mapsto d(z,y) - d(z,x),$$

has a limit when $z \in X$ tends to a point p in ∂X. This limit, denoted $\gamma_{x,y}(p)$, defines a smooth function $\gamma_{x,y}$ on ∂X, called the Busemann cocycle. The explicit formula in the above model is

$$\gamma_{x,y}(p) = \log \left| \frac{\langle y, p \rangle}{\langle x, p \rangle} \right|$$

for $x, y \in X$, $p \in \partial X$ (assuming the normalization $\langle x, x \rangle = \langle y, y \rangle = 1$).

This defines a G-invariant map $X \times X \to C^\infty(\partial X)$ satisfying the cocycle relation

$$\gamma_{x,y} + \gamma_{y,z} = \gamma_{x,z}.$$

Taking the exponential of $\gamma_{x,y}$ yields a multiplicative cocycle which admits the following interpretation.

Proposition 3.1.2. *For $x, y \in X$, the Radon–Nikodym derivative of μ_y with respect to μ_x is given by*

$$\frac{d\mu_y}{d\mu_x} = e^{-r\gamma_{x,y}}$$

where $r = k(n+1) - 2$.

The significance of the number r is that the volume of a sphere of radius ρ in X grows as $e^{r\rho}$. To prove the proposition, consider for any $x \in X$ and $t > r$, the probability measure $C^{-1} e^{-td(x,z)} \, d\mathrm{vol}(z)$ on $\overline{X} = X \cup \partial X$, where $C = \int_X e^{-td(x,z)} \, d\mathrm{vol}(z)$. When $t \to r$, it has a limit concentrated on ∂X, which is just the visual measure μ_x. Now for two base points x and y, it follows that $d\mu_y/d\mu_x$ is the limit when $t \to r$ of $\exp(-t(d(y,z) - d(x,z)))$ restricted to the boundary.

The main object we shall need in the next paragraph is the Gromov scalar product f_x.

Proposition 3.1.3. *Let x be a point in X. The function*

$$(z, w) \mapsto d(z, w) - d(z, x) - d(w, x)$$

on $X \times X$ (that is, the Gromov scalar product of $z, w \in X$ with respect to $x \in X$), has a limit when z and w tend to two distinct points p and q of ∂X. This limit, denoted $f_x(p, q)$, defines a smooth function on $\partial X \times \partial X \setminus \Delta$, where Δ denotes the diagonal. The explicit formula is

$$f_x(p, q) = \log \left| \frac{\langle p, q \rangle \langle x, x \rangle}{\langle p, x \rangle \langle q, x \rangle} \right|.$$

In particular, when $p \to q$, $f_x(p, q)$ tends to infinity like $\log | \langle p, q \rangle |$.

The map $x \mapsto f_x$ from X to $C^\infty(\partial X \times \partial X \setminus \Delta)$ is clearly G-invariant. For $x, y \in X$ and $p, q \in \partial X$,

$$f_x(p, q) - f_y(p, q) = \gamma_{x,y}(p) + \gamma_{x,y}(q).$$

3.2 Construction of a quadratic form

Let X and G be as above.

Theorem 3.2.1. *There exist a vector space E and*

(1) *a representation π of G on E,*

(2) *a map $c \colon X \times X \to E$ which is a G-equivariant cocycle, that is,*

$$c(x, y) + c(y, z) = c(x, z) \qquad and \qquad c(gx, gy) = \pi(g)c(x, y),$$

(3) *a quadratic form Q preserved by π and a function $\varphi \colon [0, \infty[\, \to \mathbb{R}$ such that $\varphi(r) \to +\infty$ when $r \to \infty$ and $Q(c(x, y)) = \varphi(d(x, y))$.*

Proof. This theorem is proved in Proposition 3.2.2, Lemma 3.2.3 and Proposition 3.2.4 below. $\qquad \square$

The space E is the space of differential forms on ∂X of maximal degree and with zero integral, and the representation π is given by the action of G on ∂X by orientation-preserving diffeomorphisms. For $x, y \in X$, we define $c(x, y)$ in E to be $\mu_y - \mu_x$. Then c is a G-equivariant cocycle.

Proposition 3.2.2. *For $\alpha \in E$, the integral*

$$Q(\alpha) = -\int_{\partial X \times \partial X} f_x(p, q)\, \alpha(p)\, \alpha(q)$$

exists and is independent of $x \in X$. It therefore defines a G-invariant quadratic form on E.

Proof. Note that the above integral makes sense because of the logarithmic behaviour of the kernel f_x on the diagonal. When x is replaced by y, the correction terms are $\int \gamma_{x,y}(z)\, \alpha(z)\, \alpha(w)$ and $\int \gamma_{x,y}(w)\, \alpha(z)\, \alpha(w)$, which are both zero because $\int \alpha = 0$. $\square$

Let S be the symmetric bilinear form associated to Q, that is, $S(\alpha, \alpha) = Q(\alpha)$. The corresponding map from E to its dual is a special case of the Knapp–Stein intertwining operators. The following lemma says that the image of the cocycle $c(x, y)$ under this map is precisely the cocycle $\gamma_{x,y}$.

Lemma 3.2.3. *For all $\beta \in E$ and $x, y \in X$,*

$$S(c((x, y), \beta)) = -\int_{\partial X} \gamma_{x,y}\, \beta.$$

Proof. Observe that

$$-S(\mu_y - \mu_x, \beta) = \int f_x(p, q)(\mu_y(p) - \mu_x(p))\, \beta(q)$$

$$= \int (f_x(p, q) - f_y(p, q))\, \mu_y(p)\, \beta(q)$$

$$+ \int f_y(p, q)\, \mu_y(p)\, \beta(q) - \int f_x(p, q)\, \mu_x(p)\, \beta(q),$$

where the integration is over the variable (p, q) in $\partial X \times \partial X$. But the last two terms are zero. Indeed, the map $q \mapsto \int f_x(p, q)\, \mu_x(p)$ is constant, because the stabilizer of x in G acts transitively on ∂X, and $\int \beta = 0$. The remaining term is the sum of $\int \gamma_{x,y}(p)\, \mu_y(p)\, \beta(q)$, which is zero because $\int \beta = 0$, and of $\int \gamma_{x,y}(q)\, \mu_y(p)\, \beta(q)$, which is equal to $\int_{\partial X} \gamma_{x,y}\beta$ since $\int \mu_y = 1$. $\square$

We are now able to compute $Q(c(x, y))$.

Proposition 3.2.4. $Q(\mu_y - \mu_x) = \varphi(d(x, y))$, *where* $\varphi(r) \to +\infty$ *when* $r \to +\infty$.

Proof. By Lemma 3.2.3,

$$Q(\mu_y - \mu_x) = S(\mu_y - \mu_x, \mu_y - \mu_x) = -\int \gamma_{x,y}\, (\mu_y - \mu_x).$$

Since the map $y \mapsto \int \gamma_{x,y} \mu_x$ is invariant under the stabilizer G_x, there exists a function ψ such that $\int \gamma_{x,y} \mu_x = \psi(d(x,y))$. Similarly $\int \gamma_{x,y} \mu_y = -\psi(d(x,y))$ and $Q(\mu_y - \mu_x) = 2\psi(d(x,y)) = \varphi(d(x,y))$. The explicit formula for φ is

$$\varphi(r) = 2 \int_S \log|\cosh r - \sinh r u_1| \, d\mathrm{vol}(u)$$

where S and $d\mathrm{vol}$ denote the unit sphere in K^n and the volume form thereon, and u_1 denotes the first coordinate of $u \in S$. Note that $\varphi(r)$ behaves like $2 \log \cosh r$ since the integral of $\log|1 - u_1|$ converges.

The proposition follows. $\qquad\square$

3.3 Positivity

Lemma 3.3.1. *If $K = \mathbb{R}$ or $\mathbb{C}$, the quadratic form Q is positive.*

Proof. It is enough to check that for any top degree form α on the unit sphere S of K^n,

$$-\int_{S \times S} \log|1 - (z,w)| \, \alpha(z)\alpha(w) \geq 0,$$

where $(z,w) = \sum \bar{z}_i w_i$. But this follows from the expansion

$$-\log|1 - (z,w)| = \sum_{k=1}^{\infty} \frac{1}{k} \mathrm{Re}\big((z,w)^k\big)$$

and the positive definiteness of the kernels $(z,w) \mapsto \mathrm{Re}\big((z,w)^k\big)$. $\qquad\square$

It is the latter fact that fails when $K = \mathbb{H}$. This is closely related to the noncommutativity of the quaternions. Indeed, the fact that the kernels $\mathrm{Re}((z,w)^k)$ are positive definite follows from the fact that

$$(z,w)^k = (z^{\otimes k}, w^{\otimes k})$$

where the kth tensor powers live in $K^n \otimes \cdots \otimes K^n$, which is defined only for commutative K.

After completion, E becomes a Hilbert space equipped with a unitary representation π and a proper G-equivariant cocycle c. This yields the desired result.

Corollary 3.3.2. *The groups $\mathrm{SO}_0(n,1)$ and $\mathrm{SU}(n,1)$ have the Haagerup property.*

More precisely, if X is a real or complex hyperbolic space, the kernel $\varphi(d(x,y))$ is conditionally negative definite. A possibly simpler proof of that result was given in [FH74]. The above approach has the advantage of being geometric and based on the construction of the quadratic form Q which is common to all rank one symmetric spaces.

On the other hand, the groups $\mathrm{Sp}(n,1)$ and $F_{4(-20)}$ have Kazhdan's property (T). In particular, the quadratic form Q has both positive and negative vectors. For an explicit study of the representation π and the quadratic form Q in the language of representation theory, see [JW77].

3.4   The link with complementary series

For $s \in [0,1]$, let E_s be the space of s-densities on ∂X. Given a choice of $x \in X$, any element ν of E_s may be written $\nu = f\, d\mu_x^s$. Note that the product of an s-density and a $(1-s)$-density is a measure, and may thus be integrated. This defines a duality between E_s and E_{1-s}.

Proposition 3.4.1. *For $s \in \,]1/2, 1[$ and $\nu \in E_s$, the integral*

$$Q_s(\nu) = \int_{\partial X \times \partial X} e^{-(1-s)rf_x(p,q)}\, (d\mu_x^{1-s}\nu)(p)\,(d\mu_x^{1-s}\nu)(q)$$

makes sense and is independent of x. It therefore defines a G-invariant quadratic form Q_s on E_s.

The condition that $s > 1/2$ is required for the local integrability of the kernel $e^{-(1-s)rf_x(p,q)}$. Proposition 3.1.2 and the remark following Proposition 3.1.3 tell us how f_x and $d\mu_x$ behave when x is replaced by y. It follows that the double $(1-s)$-density $e^{-(1-s)rf_x(p,q)}\,(d\mu_x)^{1-s}(p)\,(d\mu_x)^{1-s}(q)$ on $\partial X \times \partial X$ does not change.

When s tends to $1/2$, the form Q_s becomes the L^2-norm for half-densities. More precisely, if ν is a half-density on ∂X, then the limit of $Q_{1/2+\varepsilon}(\nu\, d\mu_x^\varepsilon)$ is $C \int \nu^2$ where C is a constant, depending on n and k.

The following proposition says that, on the other hand, when s tends to 1, then after renormalization, Q_s becomes the quadratic form Q of Proposition 3.2.2.

Proposition 3.4.2. *If α is a measure on ∂X, then*

$$Q_{1-\varepsilon}(\alpha\, d\mu_x^{-\varepsilon}) = \left(\int \alpha\right)^2 + \varepsilon Q(\alpha) + O(\varepsilon^2)$$

when $\varepsilon \to 0$.

In other words, when $s \to 1$, the space E_s equipped with Q_s degenerates to the space E_1 of measures, admitting the obvious exact sequence

$$0 \to E \to E_1 \to \mathbb{C} \to 0,$$

where E is equipped with the quadratic form Q. The above exact sequence has a (nonequivariant) lifting $1 \to \mu_x$ which produces the cocycle $c(x, y) = \mu_y - \mu_x$ with values in E considered in Lemma 3.2.3. The following result is essentially due to B. Kostant [Kos69].

Theorem 3.4.3. *If $K = \mathbb{R}$ or $\mathbb{C}$, the quadratic form Q_s is positive definite for any $s \in \ ]1/2, 1[$. If $K = \mathbb{H}$, it is positive definite if and only if $s < 1 - 1/(2n+1)$.*

When Q_s is positive definite, let π_s be the representation of G in the completion of E_s for the Hilbert norm associated to Q_s. These unitary representations are the so-called complementary series of G.

In the case of $\mathbb{R}$ or $\mathbb{C}$, the positivity of all the Q_s implies the positivity of Q by Proposition 3.4.2, and hence that G has the Haagerup property. At the same time, the complementary series representations π_s converge when $s \to 1$ to the trivial representation, thus realizing concretely the negation of property (T).

Chapter 4

Classification of Lie Groups with the Haagerup Property
by Pierre-Alain Cherix, Michael Cowling and Alain Valette

4.0   Introduction

The aim of this chapter is to establish the following classification result for connected Lie groups with the Haagerup property, already mentioned in Section 1.4.

Theorem 4.0.1. *Let G be a connected Lie group. The following conditions are equivalent:*

(1) *G has the Haagerup property;*

(2) *if a closed subgroup H is such that (G,H) has relative property (T), then H is compact;*

(3) *if a one-parameter closed subgroup H is such that (G,H) has relative property (T), then H is compact;*

(4) *G is locally isomorphic to a direct product*

$$M \times SO(n_1, 1) \times \cdots \times SO(n_k, 1) \times SU(m_1, 1) \times \cdots \times SU(m_l, 1),$$

where M is an amenable Lie group.

It is obvious that condition (1) implies condition (2), in view of the remarks made in Section 1.4. Next, it is trivial that condition (2) implies con-

The research of the first author of this chapter was supported by the Australian Research Council and the School of Mathematics of the University of New South Wales.

dition (3). Hence it remains to prove that condition (3) implies condition (4) and condition (4) implies condition (1); we call these two implications step one and step two of the classification. The proofs take up Sections 4.1 and 4.2 respectively.

4.1   Step one of the classification

4.1.1   The fine structure of Lie groups

A basic reference for this subsection is the book of G. Hochschild [Hoc65]. We assume that Lie groups are connected. If N and H are subgroups of a group G, we denote by $[H, N]$ the subgroup generated by the commutators $hnh^{-1}n^{-1}$, with $h \in H$ and $n \in N$. We note that, if H normalizes N, then $[H, N]$ is normalized both by H and N.

Lemma 4.1.1. *Let $G = RS$ be the Levi decomposition of a Lie group (that is, R is the solvable radical, S a semisimple factor). Suppose that S does not centralize R; then S does not centralize $[S, R]$ either.*

Proof. From the remark preceding the lemma, $[S, R]$ is a normal subgroup of G. Denote by $\mathfrak{r}$ the Lie algebra of R. The homomorphism Aut from R to $\mathrm{Aut}(\mathfrak{r})$ is injective ([Hoc65, IX.1.2]), so it follows from the assumption that there exists $s_0 \in S$ and $X \in \mathfrak{r}$ such that $\mathrm{Ad}(s_0)X \neq X$. For small enough positive t, it follows that

$$s_0 \exp(tX)s_0^{-1} \neq \exp(tX), \tag{4.1}$$

which we rewrite as

$$\exp(-tX)s_0 \exp(tX) \neq s_0.$$

Let S_t denote $\exp(-tX)\, S\, \exp(tX)$, the conjugate of S by $\exp(-tX)$.

We claim that there exists $t \in \mathbb{R}$ such that $S_t \neq S$. Indeed, suppose by contradiction that $S_t = S$ for all $t \in \mathbb{R}$. Let H be the subgroup generated by S and $(\exp(tX))_{t \in \mathbb{R}}$, and let $\mathfrak{h}$ be its Lie algebra. Since $\mathfrak{h}$ and $\mathfrak{r}$ are $\mathrm{Ad}(S)$-invariant, then so is $\mathfrak{h} \cap \mathfrak{r} = \mathbb{R}X$. Since S has no nontrivial multiplicative homomorphisms into $\mathbb{R}^+$, it follows that $\mathrm{Ad}(s)X = X$ for every $s \in S$, thus $s \exp(tX)s^{-1} = \exp(tX)$ for every $s \in S$ and $t \in \mathbb{R}$, which contradicts (4.1).

So take $t \in \mathbb{R}$ such that $S_t \neq S$. There exists $x \in [S, R]$ such that $xSx^{-1} = S_t$ (see [Hoc65, XI.3.3]). Since $S_t \neq S$, there exists $s_1 \in S$ such that $xs_1x^{-1} \neq s_1$, that is, $s_1^{-1}xs_1 \neq x$. Hence S does not centralize $[S, R]$. $\square$

We introduce some notation. For a G-module W, we denote by W_G the linear span of the $gX - X$, where $g \in G$ and $X \in W$; this is a G-submodule of

W. We notice that, if G is semisimple and W is a finite-dimensional G-module, then G has no nonzero fixed vectors in W_G. Indeed, by semisimplicity we may write $W = W^G \oplus W_1$, where W^G is the module of G-fixed points and W_1 is a direct sum of simple, nontrivial modules; then $W_G = W_1$ (since $V_G = V$ for a simple, nontrivial module V).

Lemma 4.1.2. *Let N and S be analytic subgroups of a Lie group, with Lie subalgebras $\mathfrak{n}$ and $\mathfrak{s}$ respectively. Assume that N is nilpotent and simply connected, that S is semisimple, and that S normalizes N. Then*

(1) *the Lie algebra of $[N, S]$ is the ideal $\mathfrak{i}$ of $\mathfrak{n}$ generated by $\mathfrak{n}_S$, where S acts on $\mathfrak{n}$ by the adjoint action Ad;*

(2) *the action of S on $[N, S]/[N, [N, S]]$ has no fixed points except the identity.*

Proof. For $s \in S$ and $X \in \mathfrak{n}$, the curve $t \mapsto s \exp(tX) s^{-1} \exp(-tX)$ is contained in $[N, S]$ and its derivative when $t = 0$ is $\mathrm{Ad}(s)X - X$; since the Lie algebra of $[N, S]$ is an ideal in $\mathfrak{n}$, it contains the ideal $\mathfrak{i}$. Conversely, for $s \in S$, $X \in \mathfrak{n}$, by the Campbell–Hausdorff formula,

$$s \exp(X) s^{-1} \exp(-X) = \exp(\mathrm{Ad}(s)X) \exp(-X)$$
$$= \exp(\mathrm{Ad}(s)X - X + \theta(\mathrm{Ad}(s)X, X)),$$

and by [Hoc65, X.3.1], $\theta(\mathrm{Ad}(s)X, X)$ lies in the smallest subspace of $\mathfrak{n}$ containing $[\mathrm{Ad}(s)X, X]$ (which is equal to $[\mathrm{Ad}(s)X - X, X]$) and stable under $\mathrm{ad}(X)$ and $\mathrm{ad}(\mathrm{Ad}(s)X)$. In particular,

$$\mathrm{Ad}(s)X - X + \theta(\mathrm{Ad}(s)X, X) \in \mathfrak{i}.$$

The analytic subgroup corresponding to the ideal $\mathfrak{i}$ therefore contains $[N, S]$, so $\mathfrak{i}$ contains the Lie algebra of $[N, S]$, proving (1).

To prove the second point, we appeal to the facts that the exponential map $\exp \colon \mathfrak{n} \to N$ is an S-equivariant diffeomorphism ([Hoc65, XII.2.1], and that the Lie algebra of $[N, [N, S]]$ is $[\mathfrak{n}, \mathfrak{i}]$ ([Hoc65, XII.3.1]). Therefore it is enough to check that $\mathfrak{i}/[\mathfrak{n}, \mathfrak{i}]$ has no nonzero fixed points under the action of S. From the first part of the lemma, we deduce that $\mathfrak{i} = \mathfrak{n}_S + [\mathfrak{n}, \mathfrak{i}]$. Since S is semisimple, the quotient module $\mathfrak{i}/[\mathfrak{n}, \mathfrak{i}]$ may be identified with a submodule of $\mathfrak{n}_S$; the result then follows from the remark preceding the lemma. $\qquad\square$

Lemma 4.1.3. *With the hypotheses and notation of Lemma 4.1.2, assume also that S centralizes $[N, N]$. Then*

(1) $[N, [N, S]] = [[N, S], [N, S]]$;

(2) *the nilpotent rank of $[N, S]$ is at most 2;*

(3) $[N, S]/[[N, S], [N, S]]$ *is a vector group on which S acts with no nonzero fixed points.*

Proof. By the remark preceding Lemma 4.1.2, $\mathfrak{n} = \mathfrak{n}^S \oplus \mathfrak{n}_S$. We begin by showing that

$$[X, Y] = 0 \qquad \forall X \in \mathfrak{n}^S \quad \forall Y \in \mathfrak{n}_S. \tag{4.2}$$

Indeed, Y is a sum of elements of the form $\mathrm{Ad}(s)Z - Z$, with $Z \in \mathfrak{n}$, and, since $X = \mathrm{Ad}(s)X$,

$$[X, \mathrm{Ad}(s)Z - Z] = [\mathrm{Ad}(s)X, \mathrm{Ad}(s)Z] - [X, Z] = \mathrm{Ad}(s)[X, Z] - [X, Z] = 0,$$

where the last equality follows from our assumption that S centralizes $[N, N]$. We now prove the lemma itself.

First, for $X \in \mathfrak{n}$, write $X = X^S + X_S$ in the decomposition $\mathfrak{n} = \mathfrak{n}^S \oplus \mathfrak{n}_S$. For $Y \in \mathfrak{n}_S$, by (4.2),

$$[X, Y] = [X_S, Y] \in [\mathfrak{i}, \mathfrak{i}].$$

Hence $[\mathfrak{n}, \mathfrak{n}_S] \subset [\mathfrak{i}, \mathfrak{i}]$. Since $\mathfrak{i}$ is the ideal generated by $\mathfrak{n}_S$ (Lemma 4.1.2(1)), and moreover $[\mathfrak{i}, \mathfrak{i}]$ is an ideal in $\mathfrak{n}$, we have $[\mathfrak{n}, \mathfrak{i}] \subseteq [\mathfrak{i}, \mathfrak{i}]$; the converse inclusion is obvious.

Second, let $\mathfrak{z}(\mathfrak{i})$ be the centre of $\mathfrak{i}$, that is, $\mathfrak{z}(\mathfrak{i})$ is the set of all $X \in \mathfrak{i}$ such that $[X, Y] = 0$ for all $Y \in \mathfrak{i}$. We want to show that $[\mathfrak{i}, \mathfrak{i}] \subseteq \mathfrak{z}(\mathfrak{i})$. Define inductively $\mathfrak{n}_0 = \mathfrak{n}_S$ and $\mathfrak{n}_j = [\mathfrak{n}, \mathfrak{n}_{j-1}]$ when $j > 0$. Since $\mathfrak{i}$ is the ideal generated by $\mathfrak{n}_S$ (Lemma 4.1.2(1)), it is enough to show that $[\mathfrak{i}, \mathfrak{i}]$ commutes with $\mathfrak{n}_j$ when $j \geq 0$; we do that by induction.

When $j = 0$, it follows from the assumptions that $[\mathfrak{i}, \mathfrak{i}] \subseteq \mathfrak{n}^S$, and $\mathfrak{n}^S$ commutes with $\mathfrak{n}_0$ by (4.2).

Assume inductively that $[[\mathfrak{i}, \mathfrak{i}], \mathfrak{n}_{j-1}] = 0$. Choose $X \in [\mathfrak{i}, \mathfrak{i}]$, $Y \in \mathfrak{n}$, and $Z \in \mathfrak{n}_{j-1}$; by the Jacobi identity,

$$[X, [Y, Z]] = [Y, [X, Z]] + [Z, [Y, X]].$$

But $[X, Z] = 0$ by the inductive hypothesis; further, $[Y, X] \in [\mathfrak{n}, \mathfrak{i}] = [\mathfrak{i}, \mathfrak{i}]$ by the first part of the lemma. So, $[Z, [Y, X]] = 0$, again by the inductive hypothesis.

The final part of the lemma follows from the equality

$$[N, [N, S]] = [[N, S], [N, S]]$$

and the fact, proved in Lemma 4.1.2(2), that S has no nonzero fixed points on $[N, S]/[N, [N, S]]$. $\qquad\square$

4.1.2   A criterion for relative property (T)

Proposition 4.1.4. *Let G be a connected Lie group. Let N and S be two analytic subgroups, with N closed, simply connected, and nilpotent, and S semisimple with no compact factors. Suppose that S normalizes N and centralizes $[N, N]$. Then the pair $(G, [N, S])$ has relative property (T).*

Proof. First notice that, since N is a closed simply connected subgroup, and $[N, S]$ is an analytic subgroup, it follows that $[N, S]$ is closed in G. Denote by $([N, S]S)^-$ the closure of $[N, S]S$ in G. It is clearly sufficient to prove that the pair $(([N, S]S)^-, [N, S])$ has relative property (T). In view of Lemma 4.1.3(2), we distinguish two cases.

First, if $[N, S]$ is abelian, we observe that $[N, S]$ is a vector group, on which S acts with no nonzero fixed points, by Lemma 4.1.3(3). By [Val94, Prop. 2.3] (which uses the assumption that S has no compact factors), the pair $(([N, S]S)^-, [N, S])$ has relative property (T).

Otherwise, $[N, S]$ is two-step nilpotent. In this case, we shall need a characterization of relative property (T) that may be added to the list in Definition 1.4.1, and appears in [HV89, p. 16]: there exists a neighbourhood V of the trivial one-dimensional representation in the unitary dual of $([N, S]S)^-$ such that every representation π in V is constant on $[N, S]$, that is, $\pi(n) = 1$ for $n \in [N, S]$. So let $(\pi_i)_{i \in I}$ be a net of irreducible unitary representations of $([N, S]S)^-$, converging to the trivial representation. We have to show that the restriction of π_i to $[N, S]$ is eventually the trivial representation on $\mathcal{H}_{\pi_i}$.

Observe that $[[N, S], [N, S]]$ is centralized by $[N, S]$ and by S, and consequently by $([N, S]S)^-$. Denote by $\overline{\pi}_i$ the contragredient representation of π_i; then the representation $\pi_i \otimes \overline{\pi}_i$ factors through $([N, S]S)^- / [[N, S], [N, S]]$. Since $[N, S]/[[N, S], [N, S]]$ is a vector group on which S acts with no nonzero fixed points (Lemma 4.1.3(3)), [Val94, Prop. 2.3] applies again to show that the pair

$$(([N, S]S)^- / [[N, S], [N, S]],\ [N, S]/[[N, S], [N, S]])$$

has relative property (T). Therefore we may assume that, for every $i \in I$, the restriction to $[N, S]$ of the representation $\pi_i \otimes \overline{\pi}_i$ contains the trivial representation. By a standard argument, this means that the restriction of π_i to $[N, S]$ contains a finite-dimensional subrepresentation. Since $[N, S]$ is solvable, by Lie's theorem every finite-dimensional unitary representation is a direct sum of unitary characters. So we find a nonzero vector ξ_i in $\mathcal{H}_{\pi_i}$ such that the one-dimensional subspace $\mathbb{C}\xi_i$ is invariant under $[N, S]$; the vector ξ_i is then fixed by $[[N, S], [N, S]]$ (since a character of a group is trivial on the derived subgroup). Since $[[N, S], [N, S]]$ is central in $([N, S]S)^-$, the set of vectors in $\mathcal{H}_{\pi_i}$ fixed under $[[N, S], [N, S]]$, is an invariant subspace of π_i. Since π_i is irreducible and this subspace contains ξ_i, we conclude that the representation π_i factors through $([N, S]S)^- / [[N, S], [N, S]]$. The fact that the pair $(([N, S]S)^- / [[N, S], [N, S]],\ [N, S]/[[N, S], [N, S]])$ has relative property (T) then ensures that, for i big enough, π_i is constant on $[N, S]/[[N, S], [N, S]]$, and the proof is complete. $\square$

Example 4.1.5. (A semidirect product with the Heisenberg group) A number of cases where Proposition 4.1.4 applies, for which $[N, S]$ is a two-step nilpotent

group, may be found in [Val94, 2.11 and 2.12]. We single out the example of
the semidirect product $H_3 \rtimes \mathrm{SL}_2(\mathbb{R})$, where $\mathrm{SL}_2(\mathbb{R})$ acts in the standard way
on the three-dimensional Heisenberg group H_3; the pair $(H_3 \rtimes \mathrm{SL}_2(\mathbb{R}), H_3)$ has
relative property (T).

4.1.3   Conclusion of step one

Let $G = RS$ be the Levi decomposition of G; we denote by S_{nc} the subgroup
of S generated by all simple, noncompact factors of S. Assume the contrary,
that is, that G is not locally isomorphic to a group of the form

$$M \times \mathrm{SO}(n_1, 1) \times \cdots \times \mathrm{SO}(n_k, 1) \times \mathrm{SU}(m_1, 1) \times \cdots \times \mathrm{SU}(m_l, 1),$$

with M amenable. There are two possibilities.

First, there may be a simple factor S_1 of S_{nc} which is not locally isomorphic to $\mathrm{SO}(n, 1)$ or $\mathrm{SU}(m, 1)$. So S_1 has property (T) (by the classification of
simple Lie groups with property (T) in of [HV89, Chap. 2]). Denote by $\overline{S}_1$ the
closure of S_1 in G; then $\overline{S}_1$ clearly has property (T), and it is not compact
(for instance, by [Hoc65, XIII.2.1]). Then the pair $(G, \overline{S}_1)$ has relative property (T). If H is any nonrelatively compact one-parameter subgroup of $\overline{S}_1$,
then (G, H) has relative property (T) as well.

Alternatively, it may happen that S_{nc} does not centralize R. In this case,
we denote the analytic subgroup RS_{nc} by H. By Lemma 4.1.1 (applied to
H), S_{nc} does not centralize $[H, R]$. Now $[H, R]$ is a normal, analytic subgroup,
which moreover is nilpotent ([Hoc65, XI.3.2]); therefore $[H, R]^-$, its closure
in G, is also nilpotent. Let T be a maximal compact subgroup in $[H, R]^-$; by
[Got50, Lem. 13], T is central in the closure $\overline{H}$ of H (in particular T is a torus).
Let $p \colon \overline{H} \to \overline{H}/T$ denote the quotient homomorphism. Then $p([H, R]^-)$ is
nilpotent and simply connected. Define inductively $N_0 = p([H, R]^-)$ and $N_i = [N_{i-1}, N_{i-1}]$ when $i > 0$. Let j be such that S_{nc} centralizes N_{j+1} but not N_j; by
Proposition 4.1.4, the pair $([N_j, S_{nc}]S_{nc}, [N_j, S_{nc}])$ has relative property (T).
Let N be $p^{-1}([N_j, S_{nc}])$, a nilpotent, closed, noncompact subgroup of G; since
T is compact, the pair (NS_{nc}, N) has relative property (T) as well. A fortiori,
the pair (G, N) has relative property (T), and if H is any nonrelatively compact
one-parameter subgroup of N, then (G, H) has relative property (T).

4.2   Step two of the classification

4.2.1   The generalized Haagerup property

Recall that a **normalized positive definite function** on a locally compact group
G is a positive definite function ϕ on G such that $\phi(e) = 1$.

We will work here with the following characterization of the Haagerup property for G: there exists a sequence (ϕ_n) of normalized positive definite functions on G such that ϕ_n is in $C_0(G)$ and $\phi_n \to 1$ locally uniformly as $n \to \infty$. The aim of this section is to prove that the Lie groups appearing in (4) of Theorem 4.0.1 have the Haagerup property.

The proof of this result involves several steps. In this subsection, we introduce the generalized Haagerup property for a pair (G, Z), where Z is a central subgroup of G, and prove some basic facts about this property. In Subsection 4.2.2, we prove that if G is amenable and Z is central in G, then (G, Z) has the generalized Haagerup property, and in Subsection 4.2.3, we show that if G is locally isomorphic to $SO(n, 1)$ or $SU(n, 1)$ and Z is central in G, then (G, Z) has the generalized Haagerup property. In the final subsection we tie these results together, using a little structure theory, and prove the missing implication in Theorem 4.0.1.

Definition 4.2.1. Suppose that G is a locally compact group and Z is a closed central subgroup of G. We say that the pair (G, Z) has the **generalized Haagerup property** if, given any compact subset K of G and any ε in $\mathbb{R}^+$, there exist a compact neighbourhood V of 1 in $\hat{Z}$ (that is, a compact set containing 1 in its interior) and a family $\{\phi_\chi : \chi \in V\}$ of normalized positive definite functions $\phi_\chi : G \to \mathbb{C}$, satisfying the following conditions:

(1) $|\phi_\chi(x) - 1| < \varepsilon$ for all x in K and χ in V;
(2) $\phi_\chi(xz) = \phi_\chi(x)\chi(z)$ for all χ in V, x in G, and z in Z;
(3) $(\chi, x) \mapsto \phi_\chi(x)$ is in $C(V \times G)$;
(4) $|\phi_\chi|$ is in $C_0(G/Z)$ for every $\chi \in V$.

We refer to these as "the condition that each ϕ_χ is near to 1", "the condition that each ϕ_χ be covariant", "the condition that the family $\{\phi_\chi\}$ be continuous", and "the condition that the functions ϕ_χ be vanishing".

The vanishing condition (4) is related to the Howe–Moore property, whose definition we recall from [IN96].

Definition 4.2.2. A locally compact group G has the **Howe–Moore property** if every unitary representation of G with no nonzero fixed vectors is a C_0-representation.

The terminology follows from the basic result of R.E. Howe and C.C. Moore ([HM79]; see also [Cow79] and [Zim84a, Thm 2.2.20]): simple noncompact algebraic groups over local fields have the Howe–Moore property. Other examples appear in [LM92]. The fact that a Howe–Moore group without property (T) must have the Haagerup property was noticed and exploited by R.J. Zimmer in [Zim84b]. For that reason, Gromov claims in [Gro88] that his first definition of a-T-menability is implicit in Zimmer's paper.

Lemma 4.2.3. *Let Z be a closed central subgroup of the locally compact group G. Let V be a neighbourhood of 1 in $\hat{Z}$, and let $\{\phi_\chi \colon \chi \in V\}$ be a family of normalized positive definite functions on G. Assume that*

(1) *G/Z has the Howe–Moore property;*

(2) *the family $\{\phi_\chi \colon \chi \in V\}$ satisfies the covariance condition (2) of Definition 4.2.1;*

(3) *for every $\chi \in V$, the representation π_χ of G associated to ϕ_χ by the Gel'fand–Naimark–Segal construction has no nontrivial finite-dimensional subrepresentations.*

Then the family $\{\phi_\chi \colon \chi \in V\}$ satisfies the vanishing condition (4) of Definition 4.2.1.

Proof. Denote by $\bar{\pi}_\chi$ the contragredient representation of π_χ. By the covariance condition (2), $\pi_\chi(z) = \chi(z)$ for $z \in Z$. Hence the representation $\pi_\chi \otimes \bar{\pi}_\chi$ factors through a representation of G/Z. Since $|\phi_\chi|^2$ is a matrix coefficient of $\pi_\chi \otimes \bar{\pi}_\chi$ and G/Z has the Howe–Moore property, to show $|\phi_\chi| \in C_0(G/Z)$ it is enough to check that $\pi_\chi \otimes \bar{\pi}_\chi$ has no nonzero fixed vectors; but, by a standard argument, this follows from the fact that π_χ has no finite-dimensional subrepresentations. $\square$

Recall that if Z_0 is a closed subgroup of the locally compact abelian group Z, and $Z_0^\perp$ denotes the subgroup of the character group $\hat{Z}$ of characters which are trivial on Z_0, then $\hat{Z_0}$ is canonically isomorphic to $\hat{Z}/Z_0^\perp$ and $(Z/Z_0)\hat{\ }$ is canonically isomorphic to $Z_0^\perp$. We investigate the generalized Haagerup property briefly.

Lemma 4.2.4. *Suppose that (G, Y) and (H, Z) have the generalized Haagerup property. Then so does $(G \times H, Y \times Z)$.*

Proof. This is simple; one takes products $\phi_\chi \otimes \psi_\omega \colon G \times H \to \mathbb{C}$ of normalized positive definite functions ϕ_χ on G and ψ_ω on H. $\square$

Lemma 4.2.5. *Suppose that H is a closed subgroup of G. Then $(H, Z \cap H)$ has the generalized Haagerup property if (G, Z) does.*

Proof. Restriction to H of a normalized positive definite function ϕ_χ on G which is near 1 and covariant (that is, ϕ_χ satisfies conditions (1) and (2)) gives a normalized positive definite function on H which satisfies these conditions with $K \cap H$, H and $H \cap Z$ in place of K, G and Z. However, a technical difficulty arises at this point. Let $\pi \colon \hat{Z} \mapsto (H \cap Z)\hat{\ }$ be the map of restriction of characters of Z to characters of $H \cap Z$; this is essentially the quotient map $\hat{Z} \to \hat{Z}/(H \cap Z)^\perp$. Then the problem is that the family of normalized positive definite functions $\{\phi_\chi|_H\}$ is not parametrized by a subset of $(H \cap Z)\hat{\,}$, as required, but by a subset of $\hat{Z}$. If π is a covering map (that is, if $(H \cap Z)^\perp$ is discrete, or equivalently if $Z/(H \cap Z)$ is compact), then by reducing the size of the compact neighbourhood V of e in $\hat{Z}$ if necessary, we may suppose that $\pi \colon V \to \pi(V)$ is one-to-one, so a homeomorphism, and then take its continuous inverse $\sigma \colon \pi(V) \to V$. In this case, setting ψ_ω equal to $\phi_{\sigma(\omega)}|_H$ yields a family $\{\psi_\omega : \omega \in \pi(V)\}$ of functions with the required properties.

In the general case, we cannot construct a continuous inverse $\sigma \colon \pi(V) \to V$, so we proceed differently. Choose a continuous function $h \colon \hat{Z} \to [0, \infty[$ which vanishes outside V such that the function $\dot{h}$ on $(Z \cap H)\hat{\,}$, given by

$$\dot{h}\big(\chi(Z \cap H)^\perp\big) = \int_{(Z \cap H)^\perp} h(\chi \upsilon)\, d\upsilon,$$

where $d\upsilon$ denotes the Haar measure on $(Z \cap H)^\perp$, is equal to 1 on a compact neighbourhood W of 1 in $(Z \cap H)\hat{\,}$, and define, for ω in W,

$$\psi_\omega(x) = \int_{(Z \cap H)^\perp} h(\chi \upsilon)\, \phi_{\chi \upsilon}(x)\, d\upsilon \qquad \forall x \in H,$$

where $\omega = \chi(Z \cap H)^\perp$. It is straightforward to check that the family of functions $\{\psi_\omega : \omega \in W\}$ has the required properties. $\qquad\square$

Lemma 4.2.6. *Suppose that N is a compact normal subgroup of G. Then (G, Z) has the generalized Haagerup property if and only if $(G/N, ZN/N)$ has the generalized Haagerup property.*

Proof. Recall that ZN/N is isomorphic to $Z/N \cap Z$. Since $N \cap Z$ is compact, the set $(N \cap Z)^\perp$ of characters of Z which are trivial on $N \cap Z$ is an open subgroup of $\hat{Z}$. If (G, Z) has the generalized Haagerup property, then, given any compact subset K of G and positive real ε, there exist a neighbourhood V of 1 in $\hat{Z}$ and a family $\{\phi_\chi : \chi \in V\}$ of normalized positive definite functions on G which satisfy the conditions of the generalized Haagerup property with KN in place of N; by taking V to be a subset of $(N \cap Z)^\perp$, we may assume that all the functions ϕ_χ are constant on cosets of $N \cap Z$ in G. By averaging these functions over N, we produce new normalized positive definite functions, still satisfying the conditions of the generalized Haagerup property, which are constant on cosets on N in G; we view these as functions on G/N, and conclude that $(G/N, ZN/N)$ has the generalized Haagerup property.

Conversely, if $(G/N, ZN/N)$ has the generalized Haagerup property, we may lift the normalized positive definite functions ϕ_χ on G/N to normalized positive definite functions on G and deduce that (G, Z) has the generalized Haagerup property. $\qquad\square$

The next lemma should be compared to Proposition 6.1.1 below.

Lemma 4.2.7. *Suppose that (G_α) is the family of compactly generated open subgroups of G. Then (G, Z) has the generalized Haagerup property if and only if $(G_\alpha, Z \cap G_\alpha)$ has the generalized Haagerup property for all α.*

Proof. If (G, Z) has the generalized Haagerup property, then $(G_\alpha, Z \cap G_\alpha)$ certainly does, by Lemma 4.2.5. Conversely, given ε in $\mathbb{R}^+$ and a compact subset K of G, there is a G_α such that $K \subseteq G_\alpha$. Assume that $(G_\alpha, Z \cap G_\alpha)$ has the generalized Haagerup property; then there is a compact neighbourhood W of 1 in $(Z \cap G_\alpha)\hat{\ }$ and a family $\{\psi_\omega : \omega \in W\}$ of normalized positive definite functions on G_α which satisfy the conditions of the generalized Haagerup property, with G_α and $Z \cap G_\alpha$ in place of G and Z. Now restriction of characters gives a map $\pi_\alpha \colon \hat{Z} \to (Z \cap G_\alpha)\hat{\ }$; since $Z \cap G_\alpha$ is open in Z, the kernel of π_α is compact. We now define $\phi_\chi \colon G \to \mathbb{C}$, for χ in $\pi_\alpha^{-1}(W)$, by the rule

$$\phi_\chi(x) = \begin{cases} \psi_{\pi_\alpha(\chi)}(x) & \text{if } x \text{ is in } G_\alpha \\ 0 & \text{otherwise.} \end{cases}$$

It is now routine to check that the family $\{\phi_\chi : \chi \in \pi_\alpha^{-1}(W)\}$ of functions on G satisfies the conditions of the generalized Haagerup property.   □

Lemma 4.2.8. *Suppose that Z_0 is a closed subgroup of Z. Then $(G/Z_0, Z/Z_0)$ has the generalized Haagerup property if (G, Z) does.*

Proof. Given a compact subset K_0 of G/Z_0 and a positive real ε, we take a compact subset K of G such that $KZ_0/Z_0 = K_0$; then there exist a compact neighbourhood V of 1 in $\hat{Z}$ and a family $\{\phi_\chi : \chi \in V\}$ of normalized positive definite functions on G which satisfy the conditions of the generalized Haagerup property. Consider the subfamily of this family of functions where χ lies in $V \cap (Z_0)^\perp$. All these functions are constant on cosets of Z_0 in G, so may be identified with functions on G/Z_0, and $V \cap (Z_0)^\perp$ is a compact neighbourhood of 1 in $(Z_0)^\perp$. It is now straightforward to check that this family of functions on G/Z_0, parametrized by $V \cap (Z_0)^\perp$, satisfies the conditions of the generalized Haagerup property with $(G_0/Z_0, Z/Z_0)$ in place of (G, Z).   □

Lemma 4.2.9. *Suppose that Z_0 and Z are commensurable, that is, $Z_0 \cap Z$ is of finite index in both Z and Z_0. Then (G, Z) has the generalized Haagerup property if and only if (G, Z_0) has the generalized Haagerup property.*

Proof. It suffices to consider the case where Z_0 is a subgroup of finite index in Z. Restriction of characters gives a map $\pi \colon \hat{Z} \to (Z_0)\hat{\ }$ with finite kernel. We can find a compact neighbourhood U of 1 in $\hat{Z}$ such that $\pi \colon U \to \pi(U)$ is one-to-one, and so has a continuous inverse σ.

If $\{\phi_\chi : \chi \in V\}$ is a family of normalized positive definite functions satisfying the conditions of the generalized Haagerup property for (G, Z), then $\{\phi_{\sigma(\omega)} : \omega \in \pi(V) \cap \pi(U)\}$ is a family of normalized positive definite functions satisfying the conditions of the generalized Haagerup property for (G, Z_0).

Conversely, suppose that (G, Z_0) has the generalized Haagerup property. Let K_1 be a (finite) set of representatives of the cosets of Z_0 in Z. Given a positive real ε and compact subset K of G, take a compact neighbourhood W of 1 in $(Z_0)\hat{\ }$ and a family $\{\psi_\omega : \omega \in W\}$ of normalized positive definite functions satisfying the conditions of the generalized Haagerup property, with Z_0, $\varepsilon/2$ and KK_1 in place of Z, ε and K. We may and shall suppose that W is contained in $\pi(U)$. For χ in $\sigma(W)$, define ϕ_χ by the formula

$$\phi_\chi(x) = \frac{1}{\operatorname{card}(K_1)} \sum_{z_1 \in K_1} \psi_{\pi(\chi)}(xz_1) \overline{\chi}(z_1),$$

where $\operatorname{card}(K_1)$ is the cardinality of K_1. Note that

$$\psi_{\pi(\chi)}(xzz_0) \overline{\chi}(zz_0) = \psi_{\pi(\chi)}(xz) \overline{\chi}(z)$$

for all x in G, z in Z, and z_0 in Z_0, so that ϕ_χ is independent of the choice of the set of representatives K_1; furthermore, for all z in Z,

$$\phi_\chi(xz) = \frac{1}{\operatorname{card}(K_1)} \sum_{z_1 \in K_1} \psi_{\pi(\chi)}(xzz_1) \overline{\chi}(zz_1) \chi(z) = \phi_\chi(x) \chi(z),$$

since $\{zz_1 : z_1 \in K_1\}$ is also a set of representatives of the cosets of Z_0 in Z. Choose now a compact neighbourhood V of 1 in $\hat{Z}$ such that $|\chi(z_1) - 1| < \varepsilon/2$, for all χ in V and all z_1 in K_1.

We claim that $\{\phi_\chi : \chi \in \sigma(W) \cap V\}$ satisfies the conditions of the generalized Haagerup property. First, it is standard that ϕ_χ is a normalized positive definite function – the proof uses the representation

$$\phi_\chi(x) = \frac{1}{\operatorname{card}(K_1)^2} \sum_{z_1, z_2 \in K_1} \psi_{\pi(\chi)}(z_1^{-1} x z_2) \chi(z_1) \overline{\chi}(z_2) \qquad \forall x \in G.$$

To see that the "near 1" condition ((1) of Definition 4.2.1) is satisfied, observe that, for all x in K,

$$|\phi_\chi(x) - 1| \le \frac{1}{\operatorname{card}(K_1)} \sum_{z_1 \in K_1} |\psi_{\pi(\chi)}(xz_1) \overline{\chi}(z_1) - 1|$$

$$\le \frac{1}{\operatorname{card}(K_1)} \sum_{z_1 \in K_1} |(\psi_{\pi(\chi)}(xz_1) - 1) \overline{\chi}(z_1)| + \frac{1}{\operatorname{card}(K_1)} \sum_{z_1 \in K_1} |\overline{\chi}(z_1) - 1|$$

$$< \varepsilon/2 + \varepsilon/2 = \varepsilon,$$

as required.   □

Lemma 4.2.10. *The pair $(G, \{e\})$ has the generalized Haagerup property if and only if G has the Haagerup property.*

Proof. This is a straightforward comparison of the two definitions.   □

Lemma 4.2.11. *Let G be a second countable, locally compact group, and let Z be a closed central subgroup. Suppose that (G, Z) has the generalized Haagerup property. Then G has the Haagerup property.*

Proof. Fix a compact subset K of G and ε in $\mathbb{R}^+$. We will produce a normalized positive definite function ϕ on G such that $|\phi(x) - 1| < \varepsilon$ for all x in K and ϕ in $C_0(G)$. Since K and ε are arbitrary, this shows that G has the Haagerup property.

The definition of ϕ is simple. Let V and $\{\phi_\chi : \chi \in V\}$ be the relatively compact neighbourhood of 1 in $\hat{Z}$ and the family of normalized positive definite functions on G whose existence is guaranteed by the definition of the generalized Haagerup property, and take v in $C_c(V)$ such that $v \geq 0$ and $\int_V v(\chi)\, d\chi = 1$, where $d\chi$ denotes Haar measure on $\hat{Z}$ restricted to V. Define

$$\phi(x) = \int_V v(\chi)\, \phi_\chi(x)\, d\chi \qquad \forall x \in G.$$

It is simple to check that ϕ is a normalized positive definite function and that $|\phi(x) - 1| < \varepsilon$ for all x in K. To show that ϕ vanishes at infinity, note first that, if $(y_n)_{n \geq 1}$ is any sequence going to infinity in G/Z, then $\lim_{n \to \infty} |\phi_\chi|\, (y_n) = 0$ by the vanishing condition (4) of Definition 4.2.1. By Lebesgue's dominated convergence theorem,

$$\lim_{n \to \infty} \int_V v(\chi)\, |\phi_\chi|\, (y_n)\, d\chi = 0.$$

Fix $\delta > 0$. Since G is second countable and $|\phi(x)| \leq \int_V v(\chi)\, |\phi_\chi|\, (x)\, d\chi$, there exists a compact set K_1 in G such that

$$|\phi(x)| \leq \delta \qquad \forall x \in G \setminus K_1 Z.$$

Now consider the function $(\chi, x) \mapsto v(\chi)\, \phi_\chi(x)$ on $V \times K_1$. This function is continuous, and we may decompose V into finitely many disjoint Borel measurable subsets V_i $(i = 1, \ldots, I)$ of measure $|V_i|$, and choose points χ_i in V_i so that

$$|v(\chi)\, \phi_\chi(x) - v(\chi_i)\, \phi_{\chi_i}(x)| < \frac{\delta}{2|V|} \qquad \forall x \in K_1 \quad \forall \chi \in V_i.$$

Now if x is in K_1 and z is in Z, we have

$$\phi(xz) = \int_V v(\chi)\, \phi_\chi(xz)\, d\chi$$

$$= \int_V v(\chi)\, \phi_\chi(x)\, \chi(z)\, d\chi$$

$$= \sum_{i=1}^{I} \int_{V_i} v(\chi_i)\, \phi_{\chi_i}(x)\, \chi(z)\, d\chi$$

$$+ \sum_{i=1}^{I} \int_{V_i} [v(\chi)\, \phi_\chi(x) - v(\chi_i)\, \phi_{\chi_i}(x)]\, \chi(z)\, d\chi.$$

Observe that, since each ϕ_{χ_i} is a normalized positive definite function,

$$\left| \sum_{i=1}^{I} \int_{V_i} v(\chi_i)\, \phi_{\chi_i}(x)\, \chi(z)\, d\chi \right| \le \sum_{i=1}^{I} |v(\chi_i)|\, \|\phi_{\chi_i}\|_\infty \left| \int_{V_i} \chi(z)\, d\chi \right|$$

$$\le \|v\|_\infty \sum_{i=1}^{I} \left| \int_{V_i} \chi(z)\, d\chi \right|,$$

and by the Riemann–Lebesgue Lemma there exists a compact subset K_2 of Z such that

$$\left| \int_{V_i} \chi(z)\, d\chi \right| < \frac{\delta}{2I\, \|v\|_\infty} \qquad \forall z \in Z \setminus K_2.$$

Thus, if z is in $Z \setminus K_2$ and x is in K_1, then

$$\left| \sum_{i=1}^{I} \int_{V_i} v(\chi_i)\, \phi_{\chi_i}(x)\, \chi(z)\, d\chi \right| < \delta/2.$$

Finally,

$$\left| \sum_{i=1}^{I} \int_{V_i} [v(\chi)\, \phi_\chi(x) - v(\chi_i)\, \phi_{\chi_i}(x)]\, \chi(z)\, d\chi \right|$$

$$\le \sum_{i=1}^{I} \int_{V_i} |v(\chi)\, \phi_\chi(x) - v(\chi_i)\, \phi_{\chi_i}(x)|\, d\chi \le \sum_{i=1}^{I} |V_i| \cdot \frac{\delta}{2|V|} = \delta/2.$$

Combining these estimates, it follows that

$$|\phi(x)| < \delta \qquad \forall x \in G \setminus K_1 K_2,$$

as required. $\square$

4.2.2 Amenable groups

In this subsection, we suppose throughout that G is an amenable locally compact group and Z is a closed central subgroup.

Proposition 4.2.12. *Let Z be a closed central subgroup of the amenable group G. Then (G, Z) has the generalized Haagerup property.*

Proof. There exists a Borel measurable section $\sigma \colon G/Z \to G$ such that $(\sigma(K))^-$ is compact for any compact subset K of G/Z (see [Keh84]).

We describe the unitary representations of G induced from the character χ of the central subgroup Z. We equip the quotient G/Z with its Haar measure, denoted by dt. Let $\mathcal{H}$ be the Hilbert space $L^2(G/Z)$, and define π_χ by the formula

$$[\pi_\chi(x)\xi](t) = \xi(x^{-1}t)\,\overline{\chi}(\zeta(x,t)) \qquad \forall x \in G \quad \forall t \in G/Z,$$

where

$$\zeta(x,t) = \sigma(x^{-1}t)^{-1}x^{-1}\sigma(t) \qquad \forall x \in G \quad \forall t \in G/Z.$$

Then π_χ is a continuous unitary representation of G on $\mathcal{H}$, and $\pi_\chi|_Z$ is a multiple of χ. In particular, π_1 is the lift to G of the regular representation of G/Z on $L^2(G/Z)$.

Consider now ξ in $\mathcal{H}$ of norm 1, and define $\phi_{\chi,\xi}$ by the rule

$$\begin{aligned}
\phi_{\chi,\xi}(x) &= \langle \pi_\chi(x)\xi, \xi \rangle \\
&= \int_{G/Z} [\pi_\chi(x)\xi](t)\,\overline{\xi}(t)\, dt \\
&= \int_{G/Z} \xi(x^{-1}t)\,\overline{\xi}(t)\,\overline{\chi}(\zeta(x,t))\, dt \qquad \forall x \in G.
\end{aligned}$$

It is easy to check that $\phi_{\chi,\xi}$ is a normalized positive definite function and satisfies the covariance condition (2) in the definition of the generalized Haagerup property. Moreover,

$$|\phi_{\chi,\xi}(x)| \le \int_{G/Z} |\xi(x^{-1}t)|\,|\xi(t)|\, dt = \phi_{1,|\xi|}(x) = \langle \pi_1(x)|\xi|, |\xi| \rangle \qquad \forall x \in G,$$

and $\phi_{1,|\xi|}$ is in $C_0(G/Z)$ because π_1 is the lift to G of the regular representation of G/Z on $L^2(G/Z)$. Thus the vanishing condition (4) of Definition 4.2.1 holds.

Before choosing the neighbourhood V and establishing conditions (1) and (3) of Definition 4.2.1, we observe that, for all $\chi, \chi' \in \hat{Z}$ and all $x, x' \in G$,

$$|\phi_{\chi,\xi}(x) - \phi_{\chi',\xi}(x')| = \left| \int_{G/Z} \xi(x^{-1}t)\,\overline{\xi}(t) \left[\overline{\chi}(\zeta(x,t)) - \overline{\chi}'(\zeta(x,t)) \right] dt \right.$$

$$\left. + \int_{G/Z} \left[\xi(x^{-1}t) - \xi(x'^{-1}t) \right] \overline{\xi}(t)\,\overline{\chi}'(\zeta(x',t))\,dt \right|$$

$$\leq \int_{G/Z} |\xi(x^{-1}t)|\,|\xi(t)|\,|\chi(\zeta(x,t)) - \chi'\zeta(x',t))|\,dt$$

$$+ \int_{G/Z} \left| \xi(x^{-1}t) - \xi(x'^{-1}t) \right| |\xi(t)|\,dt$$

$$= I_1 + I_2.$$

Now

$$I_2 \leq \left(\int_{G/Z} \left| \xi(x^{-1}t) - \xi(x'^{-1}t) \right|^2 dt \right)^{1/2} \left(\int_{G/Z} |\xi(t)|^2\,dt \right)^{1/2}$$

$$= \|\pi_1(x)\xi - \pi_1(x')\xi\|_2. \tag{4.3}$$

Moreover, for every compact subset K_1 of G/Z,

$$I_1 \leq \left(\int_{G/Z} |\xi(x^{-1}t)|^2\,dt \right)^{1/2} \left(\int_{G/Z} |\xi(t)|^2\,|\chi(\zeta(x,t)) - \chi'(\zeta(x',t))|^2\,dt \right)^{1/2}$$

$$\leq \left(\int_{K_1} |\xi(t)|^2\,|\chi(\zeta(x,t)) - \chi'(\zeta(x',t))|^2\,dt \right)^{1/2} + 2 \left(\int_{(G/Z)\backslash K_1} |\xi(t)|^2\,dt \right)^{1/2}$$

$$= I_3 + I_4. \tag{4.4}$$

Fix now a compact subset K in G and $\varepsilon > 0$. We first choose ξ in $L^2(G/Z)$ such that

$$|\phi_{1,\xi}(x) - 1| < \frac{\varepsilon}{3} \qquad \forall x \in KZ;$$

this is possible since π_1 is the lift to G of the regular representation of the amenable group G/Z. Since ξ is in $L^2(G/Z)$, by taking K_1' large enough we may arrange that $I_4 < \varepsilon/3$. If x lies in K and t lies in K_1, then $\zeta(x,t)$ lies in a relatively compact subset of Z; then

$$V = \left\{ \chi \in \hat{Z} : \left(\int_{K_1} |\xi(t)|^2\,|\chi(\zeta(x,t)) - 1|^2\,dt \right)^{1/2} < \frac{\varepsilon}{3} \right\}$$

is a neighbourhood of 1 in $\hat{Z}$. By equation (4.4) applied to K_1' and $\chi' = 1$, we have $|\phi_{\chi,\xi}(x) - 1| < \varepsilon$ for $x \in K$ and $\chi \in V$; in other words, the "near 1" condition of Definition 4.2.1 holds.

It remains to check the continuity condition (3) of Definition 4.2.1. Fix $x \in G$ and $\chi \in V$; we have to show that $|\phi_{\chi,\xi}(x) - \phi_{\chi',\xi}(x')|$ may be made small, for (x', χ') close enough to (x, χ). Since π_1 is a continuous representation, then I_2 is small when x' is in a suitable compact neighbourhood C of x. Taking K_1 big enough, we may make I_4 small, independently of x and χ. Finally, if x' lies in C and t lies in K_1, then $\zeta(x,t)$ and $\zeta(x',t)$ lie in a relatively compact subset of Z, so that, by taking χ' close enough to χ in V, we may ensure that $|\chi(\zeta(x,t)) - \chi'(\zeta(x',t))|$ is uniformly small on $C \times K_1$, whence I_3 is small. We conclude by equations (4.3) and (4.4). $\square$

4.2.3   Simple Lie groups

In this subsection, we prove that if G is a simple Lie group locally isomorphic to $\mathrm{SO}(n,1)$ or to $\mathrm{SU}(n,1)$, and Z is a central subgroup of G, then (G,Z) has the generalized Haagerup property. This will be done in two stages. First, using results of M. Flensted-Jensen on the group $\widetilde{\mathrm{SU}}(n,1)$, whose centre Z is infinite, we establish that $(\widetilde{\mathrm{SU}}(n,1), Z)$ has the generalized Haagerup property. Then we establish the analogous result for the other possible pairs (G,Z) using functorial arguments and the results of Subsection 4.2.1.

4.2.4   A covering group

We first recall some results of Flensted-Jensen [FJ77] about $\widetilde{\mathrm{SU}}(n,1)$.

Consider the group $\mathrm{U}(n,1)$ of linear transformations of $\mathbb{C}^{n+1}$ which preserve the sesquilinear form

$$[z, w] = -z_1\overline{w}_1 - \cdots - z_n\overline{w}_n + z_{n+1}\overline{w}_{n+1};$$

We write G^1 for this group. We denote the $j \times j$ identity matrix by I_j, and define the subgroups G, K^1, L, A and N of G^1 to be $\mathrm{SU}(n,1)$, $\mathrm{U}(n) \times \{1\}$, $\{\dot{l}_\theta : \theta \in \mathbb{R}\}$, $\{a_y : y \in \mathbb{R}\}$ and $\{n_{z,t} : z \in \mathbb{C}^{n-1}, t \in \mathbb{R}\}$, where $\dot{l}_\theta = e^{-i\theta}I_{n+1}$,

$$a_y = \begin{pmatrix} & & & 0 & 0 \\ & I_{n-1} & & \vdots & \vdots \\ & & & 0 & 0 \\ 0 & \cdots & 0 & \cosh y & \sinh y \\ 0 & \cdots & 0 & \sinh y & \cosh y \end{pmatrix}$$

and

$$
n_{z,t} = \begin{pmatrix}
 & & & -z_1 & z_1 \\
 & I_{n-1} & & \vdots & \vdots \\
 & & & -z_{n-1} & z_{n-1} \\
\bar{z}_1 & \cdots & \bar{z}_{n-1} & 1-\alpha & \alpha \\
\bar{z}_1 & \cdots & \bar{z}_{n-1} & -\alpha & 1+\alpha
\end{pmatrix};
$$

in the last matrix, $\alpha = \|z\|^2 / 2 - it$.

Flensted-Jensen [FJ77, Thm 1.1] shows that the map sending (k, a, n) to kan is a diffeomorphism of $K^1 \times LA \times N$ onto G^1, and that $G^1 = K^1 LA K^1$; more precisely, for every x in G^1, there exists a unique element $l_\theta a_y$ of LA such that x is in $K^1 l_\theta a_y K^1$ and $y \geq 0$. He next describes a covering group $G^{\sim 1}$ of G^1 (not the universal covering group) and a covering homomorphism $\pi \colon G^{\sim 1} \to G^1$, and defines subgroups $K^{1\sim}$, $L^\sim$, $A^\sim$ and $N^\sim$ of $G^{\sim 1}$ such that $\pi \colon K^{1\sim} \to K^1$, $\pi \colon A^\sim \to A$ and $\pi \colon N^\sim \to N$ are isomorphisms, hence we write K^1, A and N rather than $K^{1\sim}$, $A^\sim$ and $N^\sim$, while $L^\sim = \{ l_\theta : \theta \in \mathbb{R} \}$ is isomorphic to $\mathbb{R}$ and $\pi \colon l_\theta \mapsto \dot{l}_\theta$ is the usual covering of the torus by $\mathbb{R}$. Then he defines A^1 to be the abelian group $L^\sim A$. He parametrizes a typical element of the Lie algebra $\mathfrak{a}^1$ of A^1 by (θ, y) in $\mathbb{R}^2$, so that $\exp(\theta, y) = l_\theta a_y$, and defines $\mathfrak{a}^{1+}$ to be the subset $\{(\theta, y) : \theta \in \mathbb{R}, y \in [0, \infty[\}$ of $\mathfrak{a}^1$. Flensted-Jensen [FJ77, Thm 1.3] proves that the map $(k, a, n) \mapsto kan$ is also a diffeomorphism of $K^1 \times A^1 \times N$ onto $G^{\sim 1}$, and so, given x in $G^{\sim 1}$, there is a unique element $H(x)$ of $\mathfrak{a}^1$ such that x is in $K^1 \exp(H(x))N$; further, $H(x)$ depends continuously on x. He also proves that $G^{\sim 1} = K^1 A^1 K^1$. Finally, he shows that the universal covering group $\widetilde{\mathrm{SU}}(n, 1)$ of $\mathrm{SU}(n, 1)$ is the closed subgroup $\pi^{-1}(\mathrm{SU}(n, 1))$ of $G^{\sim 1}$.

The centre of the group $G^{\sim 1}$ is $L^\sim$, and $L^\sim \cap \pi^{-1}(\mathrm{SU}(n, 1))$ is the centre of $\pi^{-1}(\mathrm{SU}(n, 1))$.

By Lemma 4.2.5, in order to prove that $(\widetilde{\mathrm{SU}}(n, 1), Z)$ has the generalized Haagerup property, where Z is the centre of $\widetilde{\mathrm{SU}}(n, 1)$, it suffices to show that $(G^{\sim 1}, L^\sim)$ has the property, and this is what we shall do.

4.2.5 Spherical functions

The key to our proof is the study of spherical functions on $G^{\sim 1}$. We now summarize the relevant results of Flensted-Jensen [FJ77, pp. 70–73] about these.

There are several ways to describe the spherical functions. One approach is to consider the commutative convolution algebra $C_c^\natural(G^{\sim 1}, K^1)$ of continuous, compactly supported, K^1-bi-invariant functions on $G^{\sim 1}$. Any continuous multiplicative linear functional on this algebra may be identified with integration against a suitable K^1-bi-invariant function on $G^{\sim 1}$; these functions are the spherical functions. Alternatively, they may be seen as eigenfunctions of certain differential operators on $G^{\sim 1}$. In any case, they are known explicitly, and we now describe them.

The complex dual $\operatorname{Hom}_{\mathbb{R}}(\mathfrak{a}^1, \mathbb{C})$ of $\mathfrak{a}^1$, written $\mathfrak{a}^{1*}_{\mathbb{C}}$, is parametrized by $\mathbb{C}^2$:

$$\langle (\lambda, \mu), (\theta, y) \rangle = \lambda\theta + \mu y.$$

We abbreviate (λ, μ) by ν, and we write ρ for $(0, n)$.

The spherical functions are all of the form φ_ν, given by

$$\varphi_\nu(x) = \int_{K^1} e^{\langle i\nu - \rho, H(xk) \rangle} \, dk \qquad \forall x \in G^{\sim 1} \qquad (4.5)$$

(where dk denotes Haar measure on K^1, normalized to have total mass 1), as ν ranges over $\mathfrak{a}^{1*}_{\mathbb{C}}$, and $\varphi_{\nu_1} = \varphi_{\nu_2}$ if and only if $\lambda_1 = \lambda_2$ and $\mu_1 = \pm\mu_2$. Since they are all K^1-bi-invariant, they may be described by their restrictions to A^1, and

$$\varphi_\nu(l_\theta a_y) = e^{i\lambda\theta} \left(\cosh y\right)^\lambda \varphi_\mu^{(n-1,\lambda)}(y) \qquad \forall (\theta, y) \in \mathfrak{a}^1, \qquad (4.6)$$

where $\varphi_\mu^{(n-1,\lambda)}$ is given in terms of the hypergeometric function $_2F_1$ by the formula

$$\varphi_\mu^{(n-1,\lambda)}(y) = {}_2F_1\left(\frac{n+\lambda+i\mu}{2}, \frac{n+\lambda-i\mu}{2}, n; -(\sinh y)^2\right) \qquad \forall y \in \mathbb{R}. \quad (4.7)$$

Note in particular that $\varphi_{-i\rho} = 1$. This may be seen from the explicit formulae (4.6) and (4.7), or from the integral formula (4.5), in which the integrand is identically equal to 1.

It will be important for us that certain spherical functions are positive definite. We denote by P the compact set

$$\{(\lambda, \mu) \in \mathbb{C}^2 : i\mu \in [-n, n], \lambda \in [|\mu| - n, n - |\mu|]\}.$$

Flensted-Jensen [FJ77, Remark 2.8] states that $\varphi_{\lambda,\mu}$ is positive definite when (λ, μ) is in P, and this was confirmed by Cowling and A. Korányi [CK84].

We will need the following lemma.

Lemma 4.2.13. *The family of spherical functions $\{\varphi_\nu : \nu \in P\}$ has the following properties:*

(1) *the map $\Phi \colon P \times G^{\sim 1} \to \mathbb{C}$ given by $\Phi(\nu, x) = \varphi_\nu(x)$ is continuous;*

(2) *if (λ, μ) is in P, then $|\varphi_{\lambda,\mu}| \le \varphi_{0,\mu}$;*

(3) *if (λ, μ) is in P, x is in $G^{\sim 1}$ and θ is in $\mathbb{R}$, then $\varphi_{\lambda,\mu}(xl_\theta) = \varphi_{\lambda,\mu}(x) e^{i\lambda\theta}$;*

(4) *if η is in $(-n, n)$, then $\varphi_{0,i\eta}$ is in $C_0(G^{\sim 1}/L^{\sim})$.*

Proof. Recall that

$$\varphi_\nu(x) = \int_{K^1} e^{\langle i\nu - \rho, H(xk)\rangle}\, dk.$$

Part (1) follows from this together with Lebesgue's dominated convergence theorem.

To prove part (2), observe that, for all x in $G^{\sim 1}$,

$$|\varphi_{\lambda,\mu}(x)| \leq \int_{K^1} \left| e^{\langle i(\lambda,\mu) - (0,n), H(xk)\rangle} \right|\, dk$$

$$= \int_{K^1} e^{\langle i(0,\mu) - (0,n), H(xk)\rangle}\, dk = \varphi_{0,\mu}(x),$$

since μ is imaginary.

Recall that $L^\sim$ is the centre of $G^{\sim 1}$, whence $H(xl_\theta) = H(x) + (\theta, 0)$ for all x in $G^{\sim 1}$ and θ in $\mathbb{R}$. This result, together with the integral formula (4.5), proves part (3).

Finally, part (4) may be deduced from the integral formula (4.5) or from the explicit formulae (4.6) and (4.7). □

4.2.6 The group $\widetilde{SU}(n,1)$

We now have all the ingredients to treat the group $\widetilde{SU}(n,1)$.

Proposition 4.2.14. *Let G denote the universal covering group of $SU(n,1)$ and let Z denote the centre of G. Then (G, Z) has the generalized Haagerup property.*

Proof. As observed previously, it suffices to show that $(G^{\sim 1}, L^\sim)$ has the generalized Haagerup property.

Take a compact subset K of $G^{\sim 1}$ and ε in $\mathbb{R}^+$. By continuity (that is, Lemma 4.2.13(1)), there is a neighbourhood W of $-i\rho$ in P such that

$$|\varphi_\nu(x) - 1| < \varepsilon \qquad \forall x \in K \quad \forall \nu \in W.$$

Now fix η in $(-n, 0)$ and δ in $\mathbb{R}^+$ such that $(\lambda, i\eta)$ is in W when $|\lambda| < \delta$. Since nonconstant positive definite spherical functions of $G^{\sim 1}$ define infinite-dimensional irreducible representations of $G^{\sim 1}$, it follows from Lemma 4.2.3 that the family $V = \{\varphi_{\lambda, i\eta} : \lambda \in (-\delta, \delta)\}$ has all the required properties. □

4.2.7   The groups $SO(n,1)$ and $SU(n,1)$.

Now we complete our analysis of simple Lie groups by proving the following theorem.

Theorem 4.2.15. *Let G be a connected simple Lie group locally isomorphic to $SO(n,1)$ or $SU(n,1)$, and let Z_G denote the centre of G. Then (G, Z_G) has the generalized Haagerup property.*

Proof. Let G_1 be the quotient group $\widetilde{SU}(n,1)/Z$, where Z is the centre of $\widetilde{SU}(n,1)$; then G_1 has trivial centre. Then $(G_1, \{e\})$ has the generalized Haagerup property, from Lemma 4.2.8 and Proposition 4.2.14.

Next, suppose that G_2 is a finite cover of G_1, with centre Z_2; then (G_2, Z_2) has the generalized Haagerup property, by Lemma 4.2.6. In particular, if $G_3 = SU(n,1)$ and Z_3 is its centre, then (G_3, Z_3) has the generalized Haagerup property.

Suppose that G_4 is $SO(n,1)$ and Z_4 is its centre; from Lemma 4.2.5, (G_4, Z_4) has the generalized Haagerup property, since G_4 is a closed subgroup of G_3. Finally, if G_5 is locally isomorphic to $SO(n,1)$ and Z_5 is its centre, then (G_5, Z_5) has the generalized Haagerup property, by another application of Lemma 4.2.6. $\square$

4.2.8   Conclusion of step two

Suppose that G is a connected Lie group. Then its Lie algebra $\mathfrak{g}$ decomposes as $\mathfrak{s} \ltimes \mathfrak{r}$, where $\mathfrak{r}$ is a solvable ideal in $\mathfrak{g}$ and $\mathfrak{s}$ is semisimple, and $\mathfrak{s}$ may be written as a direct sum $\sum_i \mathfrak{s}_i$ of simple subalgebras. For the convenience of the reader, we recall the statement that we want to prove.

Theorem. *Suppose that G is a connected Lie group, and that each of the noncompact simple subalgebras $\mathfrak{s}_j$ in the decomposition of $\mathfrak{g}$ centralizes $\mathfrak{r}$ and is isomorphic to some $\mathfrak{so}(n,1)$ or $\mathfrak{su}(n,1)$. Then G has the Haagerup property.*

Proof. We may write $\mathfrak{g}$ as $\mathfrak{s}_1 \oplus \cdots \oplus \mathfrak{s}_J \oplus (\mathfrak{s}' \ltimes \mathfrak{r})$, where $\mathfrak{s}_1, \ldots, \mathfrak{s}_J$ are noncompact simple Lie algebras and $\mathfrak{s}'$ is the sum of all the compact simple subalgebras. Then the universal covering group $\widetilde{G}$ of G is a direct product $S_1 \times \cdots \times S_J \times A$ where each S_j is simple, simply connected, and locally isomorphic to $SO(n,1)$ or $SU(n,1)$, while A is amenable. The centre Z of $\widetilde{G}$ is the direct product of the centres $Z_{S_1}, \ldots, Z_{S_J}$, and Z_A of the various factors, and there is a discrete subgroup Z_0 of Z such that $G = \widetilde{G}/Z_0$. By Proposition 4.2.12, (A, Z_A) has the generalized Haagerup property; further, (S_j, Z_{S_j}) has the generalized Haagerup property by Theorem 4.2.15. By Lemma 4.2.4, $(\widetilde{G}, Z)$ has the generalized Haagerup property.

By Lemma 4.2.8, $(\widetilde{G}/Z_0, Z/Z_0)$ has the generalized Haagerup property, that is, $(G, Z/Z_0)$ has the generalized Haagerup property. By Lemma 4.2.11, we conclude that G has the Haagerup property. $\qquad\square$

Chapter 5

The Radial Haagerup Property
by Michael Cowling

5.0   Introduction

Any Lie group G may be endowed with a left-invariant Riemannian metric
in many ways, and becomes a homogeneous Riemannian manifold. Some Lie
groups, endowed with such metrics, are of at least as much interest *qua* Riemannian manifolds as *qua* Lie groups. One example of this is the collection of
groups known as harmonic NA groups or as Damek–Ricci spaces. The groups
provide a useful class of test spaces for problems in Riemannian geometry, and
in particular provide counterexamples to the so-called Lichnerowicz conjecture
that any homogeneous harmonic manifold is symmetric. Radial functions on
these groups play an important role in understanding their geometry.

It therefore seems interesting to ask what role radial functions play in
their analysis. In particular, we may ask whether these groups have the "radial
Haagerup property" (defined below).

The answer to this question, namely, "it depends", was given by B. Di
Blasio [DB97b] and by A.H. Dooley and G.K. Zhang [DZ99]; it was also implicit
in work of Cowling and Haagerup [CH89]. The first two papers employ different
methodologies, and the third is unduly complicated if one just wants to answer
this question. The purpose of this chapter is to give a (hopefully) simple and
unified approach to the radial Haagerup property for harmonic NA groups.

In the next section, we define H-type groups, then harmonic NA groups,
and review their geometry. Briefly, harmonic NA groups are extensions of
certain two-step nilpotent Lie groups by a one-dimensional dilation group. In
Section 5.2, we review spherical analysis of biradial functions on H-type groups,
and in Section 5.3 we review spherical analysis on harmonic NA groups. Fi-

© Springer Basel 2001
P.-A. Cherix et al., *Groups with the Haagerup Property*,
Modern Birkhäuser Classics, DOI 10.1007/978-3-0348-0906-1_5

nally, we examine the radial Haagerup property. We have also included an appendix on special functions since these are often used in this chapter.

To conclude our introduction, we give a formal definition of the radial Haagerup property, then state our main theorem. Given a separable locally compact group G with identity element e, with a distance function d, we call a $\mathbb{C}$-valued function f on G **radial** if there is a function $F \colon [0, \infty[\to \mathbb{C}$ such that $f(x) = F(d(x, e))$ for all x in G. We say that G, equipped with the distance d, has the **radial Haagerup property** if there exists a sequence of radial positive definite functions φ_n, vanishing at infinity, such that φ_n tends to 1 locally uniformly.

Examples of groups with the radial Haagerup property include $\mathbb{R}^n$ with the Euclidean distance, and free groups with the word length distance (Example 1.2.3).

We will prove the following theorem.

Theorem 5.0.1. *Let G be a (nondegenerate) harmonic NA group. Then G has the radial Haagerup property if and only if the centre of N is one-dimensional.*

In particular, as explained in Section 5.4 below, this implies that $\mathrm{Sp}(n, 1)$ and $F_{4,-20}$ do not have the Haagerup property.

5.1   The geometry of harmonic NA groups

Suppose that $\mathfrak{n}$ is a step-two nilpotent Lie algebra, equipped with an inner product $\langle \cdot, \cdot \rangle$, and that $\mathfrak{v}$ is the orthogonal complement of the centre $\mathfrak{z}$ of $\mathfrak{n}$. Since $\mathfrak{n}$ is step-two, if $X, Y \in \mathfrak{v}$, then $[X, Y] \in \mathfrak{z}$. For Z in $\mathfrak{z}$, define the map $J_Z \colon \mathfrak{v} \to \mathfrak{v}$ by the rule

$$\langle J_Z X, Y \rangle = \langle Z, [X, Y] \rangle \qquad \forall X, Y \in \mathfrak{v}.$$

The algebra $\mathfrak{n}$ is said to be **H-type** (or Heisenberg type) if $J_Z^2 = -|Z|^2 I_\mathfrak{v}$, where $I_\mathfrak{v}$ denotes the identity map on $\mathfrak{v}$. This condition is essentially equivalent to requiring that $\mathfrak{v}$ should be a Clifford module over $\mathfrak{z}$. The connected, simply connected nilpotent Lie group N corresponding to $\mathfrak{n}$ is called an H-type group.

These groups were introduced by A. Kaplan [Kap80], who found a fundamental solution for natural subelliptic differential operators on them. Shortly after, A. Korányi [Kor82] observed that the Iwasawa N groups associated to real rank one semisimple Lie groups are H-type groups (those associated to $\mathrm{SO}(n, 1)$ may be considered as a degenerate case). Since then, these groups and algebras have been considered from algebraic ([Rie82], [Saa96]), analytic ([Ric85]) and geometric ([Dam87b], [Dam87a], [CDKR91], [CDKR98]) viewpoints.

The Lie algebra $\mathfrak{n}$ may be enlarged, as follows. Let $\mathfrak{a}$ be $\mathbb{R}$, and take H in $\mathfrak{a} \setminus \{0\}$. Define the algebra $\mathfrak{n} \oplus \mathfrak{a}$ by the additional commutation relations

$$[sH, X] = \frac{s}{2} X \qquad \forall s \in \mathbb{R} \quad \forall X \in \mathfrak{v}$$

$$[sH, Z] = s Z \qquad \forall s \in \mathbb{R} \quad \forall Z \in \mathfrak{z}.$$

We extend the inner product of $\mathfrak{n}$ to $\mathfrak{n} \oplus \mathfrak{a}$ by requiring that H is a unit vector, orthogonal to $\mathfrak{n}$. The enlarged Lie algebra $\mathfrak{n} \oplus \mathfrak{a}$ is solvable. The associated connected, simply connected Lie group is written NA, where $N = \exp(\mathfrak{n})$ and $A = \exp(\mathfrak{a})$.

As usual, we identify $\mathfrak{n} \oplus \mathfrak{a}$ with the tangent space $T_e(NA)$ to NA at the identity e. There is a unique left-invariant Riemannian metric on NA which agrees with the inner product described above on $\mathfrak{n} \oplus \mathfrak{a}$. These groups, viewed as Riemannian manifolds, include the rank one symmetric spaces of the noncompact type, and many other nonsymmetric manifolds. Indeed, if the manifold NA is symmetric, then $\mathfrak{n}$ is degenerate or $\dim \mathfrak{z} = 1, 3$, or 7; however, there exist Clifford modules over $\mathbb{R}^q$ for all q in $\mathbb{R}$, and hence there also exist H-type groups with $\dim(\mathfrak{z})$ arbitrary.

Given any Riemannian manifold M, we may define radial functions relative to an arbitrary base point. Let Δ denote the Laplace–Beltrami operator of M. Then M is said to be **harmonic** if Δf is radial wherever f is radial, for every choice of base point. If M is homogeneous, it suffices to consider a single base point. E. Damek and F. Ricci [DR92] showed that, for any H-type group N, the corresponding group NA described above is harmonic, and thereby produced a counterexample to the Lichnerowicz conjecture.

Let $\{E_1, \ldots, E_p\}$ and $\{F_1, \ldots, F_q\}$ be orthonormal bases for $\mathfrak{v}$ and $\mathfrak{z}$ respectively. Thus $p = \dim(\mathfrak{v})$ and $q = \dim(\mathfrak{z})$; we write r for $p/4 + q/2$. For x in $\mathbb{R}^p$ and z in $\mathbb{R}^q$, we write $n(x, z)$ or just (x, z) for the element $\exp(\sum_j x_j E_j + \sum_k z_k F_k)$ of N.

We write the elements of NA in different ways at different times. Sometimes we write na, where $n \in N$ and $a \in A$. At other times, we parametrize NA by $\mathbb{R}^p \times \mathbb{R}^q \times \mathbb{R}^+$, and write (x, z, s) for the element $(x, z) \exp(\log(s)H)$. It will be convenient to denote by $D_a: N \to N$ the dilation map which, in coordinates, is given by $D_{(0,0,s)}(x, z) = (s^{1/2}s, sz)$; we sometimes write $|a|$ for the number associated to the element a of A, that is, $|(0, 0, s)| = s$.

While the geometry of harmonic NA groups is not our main preoccupation, we will give a few facts and formulae. First, the multiplication is described by the formula

$$(x, z, s)(x', z', s') = (x + s^{1/2}x', z + sz' + \frac{s^{1/2}}{2} [x, x'], ss'),$$

for all x, x' in $\mathbb{R}^p$, all z, z' in $\mathbb{R}^q$, and all s, s' in $\mathbb{R}^+$.

The geodesic distance ϱ from (x, z, s) to the identity $(0, 0, 1)$ is defined by

$$\operatorname{sech} \varrho = \left(\frac{4s}{\left(1 + s + |x|^2/4\right)^2 + |z|^2} \right)^{1/2}.$$

Given X in $\mathfrak{v}$, Z in $\mathfrak{z}$ and H in $\mathfrak{a}$, we write $\tilde{X}$, $\tilde{Z}$ and $\tilde{H}$ for the left-invariant vector fields on NA which agree with X, Z and H at the identity (we also use $\tilde{X}$ and $\tilde{Z}$ for the restrictions of these vector fields to N). We write $\Delta_{\mathfrak{v}}$ and $\Delta_{\mathfrak{z}}$ for the following differential operators:

$$\Delta_{\mathfrak{v}} = \sum_{j=1}^{p} \tilde{E}_j^2 \quad \text{and} \quad \Delta_{\mathfrak{z}} = \sum_{k=1}^{q} \tilde{F}_k^2.$$

Then the Laplace–Beltrami operator Δ of NA is given by

$$\Delta = \Delta_{\mathfrak{v}} + \Delta_{\mathfrak{z}} + \tilde{H}^2 - 2r\tilde{H}.$$

Finally, the Riemannian measures on N and on NA are left-invariant by construction. The volume element on N is just $dx\, dz$, where dx and dz are the elements of Lebesgue measure on $\mathbb{R}^p$ and $\mathbb{R}^q$. The associated Riemannian volume element on NA is $s^{-2r-1}\, dx\, dz\, ds$; this is a left-invariant Haar measure on NA. The group NA is not unimodular, and $s^{-1}dx\, dz\, ds$ is a right-invariant Haar measure element. For these facts, and much more about the geometry of harmonic NA groups, see [BTV95], [DR92] and [ADY96].

5.2 Harmonic analysis on H-type groups

Traditionally, the first steps in harmonic analysis on a locally compact group are to find its irreducible unitary representations and its Plancherel formula. For an H-type group N, it is easy to show that, when restricted to the centre $Z(N)$ of N, every irreducible unitary representation ρ is a multiple of a character of $Z(N)$. If this character is trivial, then ρ factors to an irreducible unitary representation of the Abelian group $N/Z(N)$, and hence is a character. Otherwise, there exists ζ in $\operatorname{Hom}(\mathfrak{z}, \mathbb{R}) \setminus \{0\}$ such that $\rho(0, z)$ is $\exp(i\zeta(z))I_{\mathcal{H}}$, where $I_{\mathcal{H}}$ denotes the identity map on the Hilbert space $\mathcal{H}$ of the representation ρ. In this case, ρ is trivial on the central subgroup $\{(0, z) \in Z(N) : \zeta(z) = 0\}$, and this subgroup may be factored out. Then the representation ρ is essentially a representation of a Heisenberg group with a given central character, and is unique up to unitary equivalence by the Stone–von Neumann theorem. Henceforth, we denote this representation by ρ_ζ, since it is determined by ζ. It may be described as follows. Its Hilbert space $\mathcal{H}_\zeta$ is the space $H^2(\mathfrak{v}_\zeta; w_\zeta)$ of all functions on $\mathfrak{v}$ which are holomorphic relative to the complex structure

$J_{\zeta/|\zeta|}$ and which are square integrable relative to the measure $w_\zeta \, d\mathrm{vol}$, where $d\mathrm{vol}$ is the usual element of Lebesgue measure and $w_\zeta = \exp(-|\zeta| \, |\cdot|^2/4)$. The action is given by the formula

$$\rho_\zeta(x,z)\xi(v) = \exp\left(-\frac{|\zeta|^2\,|x|^2 + 2|\zeta|\,\langle v,x\rangle + 2i\,\langle J_\zeta v,x\rangle + 4i\zeta(z)}{4}\right)\xi(x+v)$$

for all ξ in $H^2(\mathfrak{v}_\zeta; w_\zeta)$ and all x and v in $\mathfrak{v}$ and z in $\mathfrak{z}$. We remark that the set of holomorphic monomials forms an orthogonal basis for $H^2(\mathfrak{v}_\zeta; w_\zeta)$. This means that representation theory may be handled very explicitly.

However, we will not need the full strength of the unitary representation theory of N. We will instead work with the commutative algebra of "biradial" functions on N, whose harmonic analysis, via the Gel'fand theory, is more straightforward.

A function $\psi \colon N \to \mathbb{C}$ will be called *biradial* if there exists a function $\Psi \colon [0,\infty[\times [0,\infty[\to \mathbb{C}$ such that

$$\psi(x,z) = \Psi(|x|, |z|) \qquad \forall (x,z) \in N.$$

If $E(N)$ is one of the usual spaces of functions on N, then we denote by $E(N)^\sharp$ the subspace of $E(N)$ of biradial functions therein.

A subset S of N will be called biradial if its characteristic function is biradial. Let $\mathcal{M}$ be the σ-algebra of Borel measurable subsets of N, and let $\mathcal{M}^\sharp$ be the subalgebra of biradial sets. There is a conditional expectation $\mathcal{E}$ from functions on N to biradial functions on N determined by these σ-algebras; this conditional expectation is an *averaging operator* on N in the sense of Damek and Ricci [DR92]. As shown by Di Blasio [DB97a], this allows us to apply the methods of spherical harmonic analysis. The essential results of the theory of biradial functions on H-type groups are:

(1) the space $L^1(N)^\sharp$ of biradial function forms a commutative Banach algebra under convolution;

(2) the Gel'fand spectrum of this algebra may be identified with the set of bounded spherical functions; these are the bounded biradial joint eigenfunctions of the two biradial distributions $\Delta_\mathfrak{v}$ and $\Delta_\mathfrak{z}$. More precisely, integration against a bounded spherical function is a multiplicative linear functional on $L^1(N)^\sharp$;

(3) all bounded spherical functions are positive definite, and every positive definite biradial function is an integral, with a positive measure, of bounded spherical functions;

(4) there is a Plancherel formula for $L^2(N)^\sharp$.

For some particular H-type groups, Korányi [Kor82] computed explicitly the bounded spherical functions and the Plancherel measure, using the classical theory of spherical functions associated to Gel'fand pairs. In light of the results of Damek and Ricci [DR92] and Di Blasio [DB97a] mentioned above, Korányi's computations extend readily to the general case. However, these spherical functions are central to our analysis, so it is worthwhile describing then and proving the Plancherel formula in some detail.

Theorem 5.2.1. *The bounded spherical functions on N which are constant on the cosets of $Z(N)$ in N, written ψ_κ, with κ in $[0, \infty[$, are given by the formula*

$$\psi_\kappa(x, z) = j_{(p)}(\kappa \, |x|).$$

The bounded spherical functions on N which are not constant on the cosets of $Z(N)$ in N, written $\psi_{h,\eta}$ where $h \in \mathbb{N}$ and $\eta \in \mathbb{R}^+$, are given by the formula

$$\psi_{h,\eta}(x, z) = \binom{h+\alpha}{h}^{-1} \exp\left(- \frac{\eta \, |x|^2}{4}\right) L_h^{(\alpha)}\left(\frac{\eta \, |x|^2}{2}\right) j_{(q)}(\eta \, |z|), \tag{5.1}$$

where $\alpha = p/2 - 1$. The functions $L_h^{(\alpha)}$ and $j_{(n)}$ are the generalized Laguerre polynomial of degree h and the modified Bessel function described in Section 5.5 below. The Plancherel formula is

$$\|f\|_2^2 = \frac{1}{2^{2r-1}\,\Gamma(q/2)\,\pi^{p+q/2}} \int_{\mathbb{R}^+} \sum_{h \in \mathbb{N}} |\hat{f}(h, \eta)|^2 \binom{h+\alpha}{h} \eta^{2r-1} d\eta$$

for all f in $L^2(N)^\sharp$.

Proof. The bounded spherical functions are joint eigenfunctions of $\Delta_\mathfrak{v}$ and $\Delta_\mathfrak{z}$. For a biradial function ψ,

$$\Delta_\mathfrak{v} \psi(x, z) = \sum_{j=1}^p \frac{\partial^2}{\partial x_j^2} \, \psi(x, z) + \frac{|x|^2}{4} \sum_{k=1}^q \frac{\partial^2}{\partial z_k^2} \, \psi(x, z),$$

while

$$\Delta_\mathfrak{z} \psi(x, z) = \sum_{k=1}^q \frac{\partial^2}{\partial z_k^2} \, \psi(x, z),$$

for all $(x, z) \in N$. Since $\Delta_z \psi = E_\mathfrak{z} \psi$, it follows that, for each x in $\mathbb{R}^p$, the function $\psi(x, \cdot)$ is a bounded radial eigenfunction of the Laplacian Δ_z on $\mathbb{R}^q$. Hence there exist η in $[0, \infty[$ and a function $F \colon [0, \infty[\to \mathbb{C}$ such that $E_\mathfrak{z} = -\eta^2$ and $\psi(x, z) = F(|x|) j_{(q)}(\eta z)$ for all (x, z) in N. We may extend F to a smooth even function on $\mathbb{R}$: indeed, if we fix an arbitrary unit vector E in $\mathbb{R}^p$, then we may define $F(t)$ to be $\psi(tE, 0)$.

On the one hand, if $\eta = 0$, then ψ is independent of the central variable, and therefore satisfies the equation

$$\sum_{j=1}^{p} \frac{\partial^2}{\partial x_j^2}\, \psi(x,z) = E_\mathfrak{v}\,\psi(x,z) \qquad \forall (x,z) \in N;$$

in this case there exists κ in $[0,\infty[$ such that $E_\mathfrak{v} = -\kappa^2$ and $\psi(x,z) = j_{(p)}(\kappa x)$ for all (x,z) in N.

On the other hand, if $\eta > 0$, then the function F satisfies the equation

$$\left(\frac{\partial^2}{\partial s^2} + \frac{p-1}{s}\frac{\partial}{\partial s} \right) F(s) - \frac{\eta^2\, s^2}{4}\, F(s) = E_\mathfrak{v}\, F(s) \qquad \forall s \in [0,\infty[;$$

the bounded solutions of this equation are multiples of

$$L_h^{(\alpha)}\left(\frac{\eta\, s^2}{2} \right) \exp\!\left(-\frac{\eta\, s^2}{4} \right)$$

(for which $E_\mathfrak{v} = -\eta(2h + p/2)$), and the formula for $\psi_{h,\eta}$ follows.

To compute the Plancherel measure for the Gel'fand transformation, we take a smooth rapidly vanishing biradial function f on N and write

$$\mathcal{F}(|x|^2, \eta) = \int_{\mathbb{R}^q} f(x,z)\, j_{(q)}(\eta z)\, dz.$$

The Gel'fand transform of f is written $\hat{f}$; $\hat{f}(h,\eta)$ is equal to

$$\int_{\mathbb{R}^p} \int_{\mathbb{R}^q} f(x,z) \binom{h+\alpha}{h}^{-1} \exp\!\left(-\frac{\eta\,|x|^2}{4} \right) L_h^{(\alpha)}\!\left(\frac{\eta\,|x|^2}{2} \right) j_{(q)}(\eta z)\, dz\, dx$$

$$= \omega_{p-1} \binom{h+\alpha}{h}^{-1} \int_{\mathbb{R}+} \mathcal{F}(s^2, \eta) \exp\!\left(-\frac{\eta\, s^2}{4} \right) L_h^{(\alpha)}\!\left(\frac{\eta\, s^2}{2} \right) s^{p-1}\, ds$$

$$= \frac{\omega_{p-1}}{2} \binom{h+\alpha}{h}^{-1} \left(\frac{2}{\eta} \right)^{p/2} \int_{\mathbb{R}+} \mathcal{F}\!\left(\frac{2t}{\eta}, \eta \right) \exp\!\left(-\frac{t}{2} \right) L_h^{(\alpha)}(t)\, t^\alpha\, dt,$$

where ω_{p-1} is the volume of S^{p-1}. By the Plancherel theorem in $\mathbb{R}^q$, if $g\colon \mathbb{R}^q \to \mathbb{C}$ is smooth, rapidly vanishing, and radial, then

$$\int_{\mathbb{R}^q} |g(z)|^2\, dz = \frac{\omega_{q-1}}{(2\pi)^q} \int_{\mathbb{R}+} |\tilde{g}(\eta)|^2\, \eta^{q-1}\, d\eta,$$

where

$$\tilde{g}(\eta) = \int_{\mathbb{R}^q} g(z)\, j_{(q)}(\eta z)\, dz.$$

If follows that, for f as above,

$$\|f\|_2^2 = \int_{\mathbb{R}^p} \int_{\mathbb{R}^q} |f(x,z)|^2 \, dz \, dx$$

$$= \int_{\mathbb{R}^p} \frac{\omega_{q-1}}{(2\pi)^q} \int_{\mathbb{R}^+} \left| \mathcal{F}(|x|^2, \eta) \right|^2 \eta^{q-1} \, d\eta \, dx$$

$$= \frac{\omega_{p-1} \omega_{q-1}}{(2\pi)^q} \int_{\mathbb{R}^+} \int_{\mathbb{R}^+} \left| \mathcal{F}(s^2, \eta) \right|^2 s^{p-1} \, ds \, \eta^{q-1} \, d\eta$$

$$= \frac{\omega_{p-1} \omega_{q-1}}{2(2\pi)^q} \int_{\mathbb{R}^+} \left(\frac{2}{\eta} \right)^{p/2} \int_{\mathbb{R}^+} \left| \mathcal{F}\left(\frac{2t}{\eta}, \eta \right) \right|^2 t^\alpha \, dt \, \eta^{q-1} \, d\eta.$$

Now the functions $t \mapsto \exp(-t/2) L_h^{(\alpha)}(t)$, where $h \in \mathbb{N}$, form an orthogonal basis for the weighted space $L^2(\mathbb{R}^+; w)$, where $w(t) = t^\alpha$ (see (5.17) below).

Thus, by the Parseval theorem,

$$\int_{\mathbb{R}^+} \left| \mathcal{F}\left(\frac{2t}{\eta}, \eta \right) \right|^2 t^\alpha \, dt$$

$$= \sum_{h \in \mathbb{N}} \frac{\left| \int_{\mathbb{R}^+} F(2t/\eta, \eta) \exp(-t/2) L_h^{(\alpha)}(t) \, t^\alpha \, dt \right|^2}{\int_{\mathbb{R}^+} \left| \exp(-t/2) L_h^{(\alpha)}(t) \right|^2 t^\alpha \, dt}$$

$$= \sum_{h \in \mathbb{N}} \frac{(\eta/2)^p \, (2/\omega_{p-1})^2 \, \binom{h+\alpha}{h}^2 \, \left| \hat{f}(h, \eta) \right|^2}{(h+\alpha)!/h!}$$

$$= \sum_{h \in \mathbb{N}} \left(\frac{\eta}{2} \right)^p \left(\frac{2}{\omega_{p-1}} \right)^2 \frac{1}{\alpha!} \binom{h+\alpha}{h} \left| \hat{f}(h, \eta) \right|^2.$$

By combining the last two formulae, we conclude that $\|f\|_2^2$ is equal to

$$\frac{\omega_{p-1} \omega_{q-1}}{2(2\pi)^q} \int_{\mathbb{R}^+} \left(\frac{2}{\eta} \right)^{p/2} \sum_{h \in \mathbb{N}} \left(\frac{\eta}{2} \right)^p \left(\frac{2}{\omega_{p-1}} \right)^2 \frac{1}{\alpha!} \binom{h+\alpha}{h} \left| \hat{f}(h, \eta) \right|^2 \eta^{2r-1} \, d\eta$$

$$= \frac{\omega_{q-1}}{2^{\alpha+q} \pi^q \omega_{p-1} \alpha!} \int_{\mathbb{R}^+} \sum_{h \in \mathbb{N}} \binom{h+\alpha}{h} \left| \hat{f}(h, \eta) \right|^2 \eta^{2r-1} \, d\eta$$

$$= \frac{1}{2^{2r-1} \Gamma(q/2) \pi^{(p+q)/2}} \int_{\mathbb{R}^+} \sum_{h \in \mathbb{N}} \binom{h+\alpha}{h} \left| \hat{f}(h, \eta) \right|^2 \eta^{2r-1} \, d\eta,$$

as required. $\square$

Remark. It may be shown that, if f is biradial, and ρ_ζ is the irreducible unitary representation of N such that $\rho_\zeta(0, z) = \exp(-i\zeta(z))I_{\mathcal{H}_\zeta}$, where ζ is in $\text{Hom}(\mathfrak{z}, \mathbb{R}) \setminus \{0\}$, then $\rho_\zeta(f)$ is diagonal in the natural monomial basis of $\mathcal{H}_\zeta$, and that $\rho_\zeta(f)$ acts on all monomials of degree h by the scalar $\hat{f}(h, |\zeta|)$. This is a measure of the degree of redundancy involved when using representation theory to deal with biradial functions.

We conclude this section by introducing some operators on $L^2(N)^\sharp$, namely the projections P_h and the complex powers of $-\Delta_{\mathfrak{z}}$. These operators may be defined on $L^2(N)$ with a little more work, but we only need them on $L^2(N)^\sharp$. For h in $\mathbb{N}$, we write P_h for the projection on $L^2(N)^\sharp$ such that

$$(P_h f)\hat{\ }(h', \eta) = \begin{cases} \hat{f}(h, \eta) & \text{if } h' = h \\ 0 & \text{otherwise.} \end{cases}$$

For λ in $\mathbb{C}$, we define the unbounded operator $(-\Delta_\mathfrak{v})^\lambda$ on $L^2(N)^\sharp$ by the formula

$$((-\Delta_\mathfrak{v})^\lambda f)\hat{\ }(h, \eta) = \eta^\lambda (2h + p/2)^\lambda \hat{f}(h, \eta).$$

This is the natural extension of $(-\Delta_\mathfrak{v})^m$ with m in $\mathbb{N}$; indeed, in this case the formula holds by integration by parts.

It follows from the definition of $\psi_{h,\eta}$ that $\psi_{h,\eta} \circ D_a = \psi_{h,|a|\eta}$ for all a in A. Hence

$$\begin{aligned}
(f \circ D_a)\hat{\ }(h, \eta) &= \int_N f(|a|^{1/2} x, |a| z)\, \psi_{h,\eta}(x, z)\, dn(x, z) \\
&= |a|^{-2r} \int_N f(x, z)\, \psi_{h,\eta}(|a|^{1/2} x, |a|^{-1} z)\, dn(x, z) \\
&= |a|^{-2r}\, \hat{f}(h, |a|^{-1} \eta).
\end{aligned}$$

From this we deduce that

$$P_h(f \circ D_a) = (P_h f) \circ D_a \qquad \forall f \in L^2(N)^\sharp \quad \forall h \in \mathbb{N} \quad \forall a \in A. \qquad (5.2)$$

It is also easy to check that

$$P_h(f_1 * f_2) = (P_h f_1) * (P_h f_2) \qquad \text{and} \qquad P_h(\overline{f}) = (P_h f)\bar{\ }. \qquad (5.3)$$

Observe also that, if $f_1, f_2 \in L^2(N)^\sharp$, then $\sum_{h\in\mathbb{N}} P_h(f_1 * f_2)$ converges absolutely; indeed

$$\sum_{h\in\mathbb{N}} \|P_h(f_1 * f_2)\|_\infty = \sum_{h\in\mathbb{N}} \|(P_h f_1) * (P_h f_2)\|_\infty$$

$$\leq \sum_{h\in\mathbb{N}} \|P_h f_1\|_2 \, \|P_h f_2\|_2$$

$$\leq \left(\sum_{h \in \mathbb{N}} \|P_h f_1\|_2^2 \right)^{1/2} \left(\sum_{h \in \mathbb{N}} \|P_h f_1\|_2^2 \right)^{1/2}$$

$$= \|f_1\|_2 \, \|f_2\|_2 . \tag{5.4}$$

Similarly, we may check that

$$(-\Delta_v)^\lambda (f \circ D_a) = |a|^\lambda \left((-\Delta_v)^\lambda f \right) \circ D_a \tag{5.5}$$

$$\left((-\Delta_v)^\lambda f \right)^{\widetilde{}} = (-\Delta_v)^{\overline{\lambda}} \widetilde{f} \tag{5.6}$$

$$(-\Delta_v)^\lambda f_1 * (-\Delta_v)^\mu f_2 = (-\Delta_v)^{\lambda+\mu} (f_1 * f_2) \tag{5.7}$$

for f, f_1 and f_2 in appropriate subsets of $L^2(N)^\sharp$, so that everything is well defined. Finally, it is clear that $(-\Delta_v)^\lambda$ and P_h is commute.

5.3   Analysis on harmonic NA groups

The irreducible unitary representations of NA fall into several categories: those which are trivial on N (these are essentially characters of A), those which are nontrivial on N but trivial on its centre $Z(N)$, and these whose restrictions to $Z(N)$ are nontrivial (these are, when restricted to N, direct integrals of the representations ρ_ζ described earlier). The only representations which are needed for the Plancherel formula are those of the third category.

The irreducible unitary representations of NA and the Plancherel formula were used in [CH89], but we will not need them here, so we give no further details. However, we will need a family of representations of NA which arise from the natural action of NA on N:

$$na \cdot n' = n \, D_a n' \qquad \forall a \in A \quad \forall n, n' \in N.$$

The Radon–Nikodym derivative of this action is $|a|^{2r}$. For λ in $\mathbb{C}$, we define the representation π_λ of NA on the space of all functions on N by the formula

$$[\pi_\lambda(na)f](n') = |a|^{\lambda-r} f \left((na)^{-1} \cdot n' \right) \qquad \forall a \in A \quad \forall n, n' \in N.$$

Observe that if $\lambda \in i\mathbb{R}$, then

$$\int_N \left| [\pi_\lambda(na)f](n') \right|^2 dn' = |a|^{-2r} \int_N \left| f \circ D_{a^{-1}}(n^{-1}n') \right|^2 dn'$$

$$= |a|^{-2r} \int_N \left| f \circ D_{a^{-1}}(n') \right|^2 dn'$$

$$= \int_N \left| f(n') \right|^2 dn',$$

that is, the representation π_λ acts unitarily on $L^2(N)$. More generally, if $\operatorname{Re}\lambda$ is in $[-r, r]$, then π_λ acts isometrically on $L^s(N)$, where $s = 2r/(r - \operatorname{Re}\lambda)$.

Lemma 5.3.1. *For all f_1, f_2 in $L^2(N)^\sharp$, and all λ in $i\mathbb{R}$,*

$$\langle \pi_\lambda(na)f_1, f_2 \rangle = |a|^{\lambda - r} \overline{f_2} * (f_1 \circ D_{a^{-1}})^{\check{}}(n), \qquad (5.8)$$

for all $a \in A$ and $n \in N$, where $\check{f}(n) = f(n^{-1})$. Consequently,

$$\langle \pi_0(\cdot)f_1, f_2 \rangle = \sum_{h \in \mathbb{N}} \langle \pi_0(\cdot)P_h f_1, P_h f_2 \rangle \qquad (5.9)$$

(the sum converges uniformly) and

$$\langle \pi_0(\cdot)(-\Delta_v)^\lambda f_1, (-\Delta_v)^{-\overline{\lambda}} f_2 \rangle = \langle \pi_\lambda(\cdot)f_1, f_2 \rangle. \qquad (5.10)$$

Proof. The first formula holds for all f_1, f_2 in $L^2(N)$. Indeed,

$$\begin{aligned}
\langle \pi_\lambda(na)f_1, f_2 \rangle &= \langle \pi_\lambda(n)\,\pi_\lambda(a)f_1, f_2 \rangle \\
&= |a|^{\lambda - r} \langle \pi_\lambda(n)(f_1 \circ D_{a^{-1}}), f_2 \rangle \\
&= |a|^{\lambda - r} \overline{f_2} * (f_1 \circ D_{a^{-1}})^{\check{}}(n).
\end{aligned}$$

When $f_1, f_2 \in L^2(N)^\sharp$, then $f_1 \circ D_{a^{-1}} \in L^2(N)^\sharp$ and $(f_1 \circ D_{a^{-1}})^{\check{}} = f_1 \circ D_{a^{-1}}$. In any case, for each a in A, $\langle \pi_\lambda(\cdot\, a)f_1, f_2 \rangle$ is in $C_0(N)^\sharp$.

By (5.2)–(5.4), we see that

$$\begin{aligned}
\langle \pi_0(\cdot\, a)f_1, f_2 \rangle &= |a|^{-r} \overline{f_2} * (f_1 \circ D_{a^{-1}})^{\check{}} \\
&= \sum_{h \in \mathbb{N}} |a|^{-r} P_h \big(\overline{f_2} * (f_1 \circ D_{a^{-1}})^{\check{}} \big) \\
&= \sum_{h \in \mathbb{N}} |a|^{-r} (P_h f_2)^{\check{}} * \big((P_h f_1) \circ D_{a^{-1}} \big)^{\check{}} \\
&= \sum_{h \in \mathbb{N}} \langle \pi_0(\cdot\, a)P_h f_1, P_h f_2 \rangle, \qquad (5.11)
\end{aligned}$$

and the convergence is uniform. Finally, by (5.5)–(5.7),

$$\begin{aligned}
\langle \pi_0(\cdot\, a)(-\Delta_v)^\lambda f_1, (-\Delta_v)^{-\overline{\lambda}} f_2 \rangle &= |a|^{-r} \big((-\Delta_v)^{-\overline{\lambda}} f_2 \big)^{\check{}} * \big((-\Delta_v)^\lambda f_1 \big) \circ D_{a^{-1}} \\
&= |a|^{\lambda - r} (-\Delta_v)^{-\lambda} \overline{f_2} * (-\Delta_v)^\lambda (f_1 \circ D_{a^{-1}}) \\
&= |a|^{\lambda - r} \overline{f_2} * (f_1 \circ D_{a^{-1}}) \\
&= \langle \pi_\lambda(\cdot\, a)f_1, f_2 \rangle.
\end{aligned}$$

Note that these formulae continue to hold for some λ which are not purely imaginary, provided that f_1 and f_2 are suitably restricted. $\qquad \square$

We now leave representation theory and take up spherical harmonic analysis. A function $\varphi\colon NA \to \mathbb{C}$ will be called *radial* if there exists a function $\Phi\colon [0, \infty[\to \mathbb{C}$ such that

$$\varphi(na) = \Phi\left(d(na, e)\right) \qquad \forall na \in NA.$$

The natural conditional expectations (also denoted $\mathcal{E}$) from functions on NA to radial functions on NA is an averaging operator in the sense of Damek and Ricci [DR92]. In particular, this means that

$$\mathcal{E}(f_1 * \mathcal{E}f_2) = \mathcal{E}f_1 * \mathcal{E}f_2 \qquad \forall f_1, f_2 \in C_c^\infty(NA).$$

(Interestingly enough, all the proofs of this fact to date use more than just geometry and measure theory.) Further, the Laplace–Beltrami operator Δ is radial, in the sense that $\mathcal{E}(\Delta f) = \Delta(\mathcal{E}f)$ for all f in $C_c^\infty(NA)$. By results of Di Blasio [DB97a], we may apply the methods of spherical harmonic analysis. The theory of spherical functions on harmonic NA groups may be found in [DR92] and [DB97a], and we merely summarize the main results:

(1) the space $L^1(NA)^\sharp$ of radial functions in $L^1(NA)$ forms a commutative Banach algebra under convolution;

(2) the Gel'fand spectrum of the algebra $L^1(NA)^\sharp$ may be identified with the set of bounded spherical functions; these are the bounded radial eigenfunctions of the radial distribution Δ; more precisely, integration against a bounded spherical function is a multiplicative linear functional on $L^1(AN)^\sharp$;

(3) only some of the bounded spherical functions are positive definite. Every positive definite radial function is an integral, with a positive measure, of positive definite spherical functions;

(4) there is a Plancherel formula for $L^2(NA)^\sharp$, involving only some of the positive definite spherical functions.

In order to be more precise, we give a description of the spherical functions, following Damek and Ricci [DR92] and J.-Ph. Anker, Damek, and C. Yacoub [ADY96], with some minor notational changes. First of all, they arise by averaging exponentials:

$$\varphi_\lambda = \mathcal{E}\left((x, z, t) \mapsto t^{r-\lambda}\right).$$

They are eigenfunctions of Δ,

$$\Delta\varphi_\lambda = (\lambda^2 - r^2)\varphi_\lambda,$$

and $\varphi_\lambda = \varphi_\mu$ if and only if $\lambda = \pm\mu$. The radial part of the Laplace–Beltrami operator may be written as

$$\frac{\partial^2}{\partial\varrho^2} + \left(\frac{p+q}{2} \coth \frac{\varrho}{2} + \frac{q}{2} \tanh \frac{\varrho}{2}\right) \frac{\partial}{\partial\varrho},$$

from which it follows that the spherical functions are essentially Jacobi functions. When $\operatorname{Re}\lambda > 0$, then Φ_λ has an asymptotic expansion (where $\varphi_\lambda(x) = \Phi_\lambda(d(x,e))$), and

$$\Phi_\lambda(\varrho) \sim \mathbf{c}(\lambda)\, e^{(\lambda-r)\varrho} \quad \text{as} \quad \varrho \to \infty;$$

when $\operatorname{Re}\lambda = 0$, there are two leading terms in the expansion. The $\mathbf{c}$-function is given by

$$\mathbf{c}(\lambda) = \frac{2^{2r-2\lambda}\,\Gamma(r+r_0)\,\Gamma(2\lambda)}{\Gamma(r+\lambda)\,\Gamma(r_0+\lambda)}.$$

The inversion formula states that, for "nice" radial functions f,

$$f(na) = \int_{\mathbb{R}^+} \hat{f}(i\lambda)\, \varphi_{i\lambda}(na)\, \frac{2^{q-2}\,\Gamma(r+r_0)}{\pi^{r+r_0+1}}\, |\mathbf{c}(\lambda)|^{-2}\, d\lambda.$$

In light of result (3) of the theory of spherical functions above, it is obviously important to know which spherical functions are positive definite. If $\lambda \in i\mathbb{R}$, then

$$\langle \pi_\lambda(\cdot)u_\lambda, u_{-\overline{\lambda}}\rangle = \langle \pi_\lambda(\cdot)u_\lambda, u_\lambda\rangle,$$

so that ϕ_λ is obviously positive definite. These functions appear in the Plancherel formula. In the next section, we shall show that φ_λ is positive definite for some other values of λ.

5.4 Positive definite spherical functions

We are finally in a position to describe the positive definite spherical functions on NA. If φ is positive definite, then $\varphi(na) = \overline{\varphi}\left((na)^{-1}\right)$, so a radial positive definite function φ is real-valued. From the eigenfunction equation, φ_λ is real-valued only if $\lambda \in \mathbb{R} \cup i\mathbb{R}$. Since positive definite functions are bounded, the only candidates for positive definite spherical functions φ_λ with λ real are those where $\lambda \in [-r, r]$. In fact, Di Blasio [DB97b] showed that φ_λ is positive definite when $\lambda \in [-r_0, r_0] \cup \{\pm r\}$, and A.H. Dooley and G.K. Zhang [DZ99] proved the converse. We shall present a unified approach to both these results, using spherical harmonic analysis of biradial functions on N and an integral representation for the spherical functions.

To examine whether φ_λ is positive definite for real values of λ, we realize φ_λ as a matrix coefficient of π_0. This will involve interpreting $(-\Delta_\mathfrak{v})^\lambda u_\lambda$ as an element of $L^2(N)$ when $\operatorname{Re}\lambda \neq 0$, which will take up the first part of this section. Then, we relate $(-\Delta_\mathfrak{v})^\lambda u_\lambda$ and $(-\Delta_\mathfrak{v})^{-\overline{\lambda}} u_{-\overline{\lambda}}$, which will take up the second part of the section. Finally, we draw conclusions from our analysis.

We need one more definition. We define the function $u_\lambda \colon N \to \mathbb{C}$ by

$$u_\lambda(x,z) = \left(\left(1 + \frac{|x|^2}{4}\right)^2 + |z|^2\right)^{\lambda - r}. \tag{5.12}$$

Then, according to Anker, Damek, and Yacoub [ADY96],

$$\varphi_\lambda = \frac{2^{q-1}\,\Gamma(r + r_0)}{\pi^{r + r_0}} \, \langle \pi_\lambda(\cdot)u_\lambda, u_{-\overline{\lambda}}\rangle. \tag{5.13}$$

Let S denote the strip $\{\lambda \in \mathbb{C} : \operatorname{Re}\lambda \in (-r, r)\}$.

Lemma 5.4.1. *The map* $\lambda \mapsto (-\Delta_\mathfrak{v})^\lambda u_\lambda$ *is analytic from* S *into* $L^2(N)$.

Proof. By induction, if $\mu \in \mathbb{N}$, then

$$(-\Delta_\mathfrak{v})^\mu u_\lambda = P_{\mu,\lambda}\, u_{\lambda - 2\mu}$$

for all λ in $\mathbb{C}$, where $P_{\mu,\lambda}$ is a polynomial – more precisely,

$$P_{\mu,\lambda}(x,z) = \sum_{j,k \in \mathbb{N}} a_{\mu,\lambda}^{j,k}\, |x|^{2j}\, |z|^{2k}$$

where $a_{\mu,\lambda}^{j,k} = 0$ unless $2j + 4k \leq 6\mu$. Thus

$$|(-\Delta_\mathfrak{v})^\mu u_\lambda(x,z)| \leq C_{\mu,\lambda}\left(\left(1 + \frac{|x|^2}{4}\right)^2 + |z|^2\right)^{3\mu/2} |u_{\lambda - 2\mu}|$$

$$= C_{\mu,\lambda}\, u_{\operatorname{Re}\lambda - \mu/2}.$$

Moreover, if $\operatorname{Re}\nu \in {]0, 2r[}$, then there exists a locally integrable function K_ν on N, homogeneous and smooth in $N \setminus \{(0,0)\}$ (this will be written N^* below), depending analytically on ν, such that

$$|K_\nu(x,z)| \leq C_\nu\bigl(|x|^4 + |z|^2\bigr)^{\operatorname{Re}\nu/2 - r} \qquad \forall (x,z) \in N^*$$

and

$$(-\Delta_\mathfrak{v})^{-\nu} f = K_\nu * f$$

for smooth functions f which vanish fast enough at infinity for the integral defining the convolution to converge absolutely at every point of N. For more details, see, for instance, [Fol75]. In particular, if $\mu \in \mathbb{N}$ and $\lambda, \nu \in \mathbb{C}$, and

$\operatorname{Re}\nu/2 + \operatorname{Re}\lambda - \mu/2 < r$, then the integral defining $K_\nu * (-\Delta_\mathfrak{v})^\mu u_\lambda$ converges absolutely at every point and is bounded in N. Further, if also $\operatorname{Re}\lambda - \mu \in \ ]0, 2r[$, then

$$
\begin{aligned}
|K_\nu * (-\Delta_\mathfrak{v})^\mu u_\lambda| &\leq |K_\nu| * |(-\Delta_\mathfrak{v})^\mu u_\lambda| \\
&\leq C_{\lambda,\mu,\nu}\, K_{\operatorname{Re}\nu} * K_{2\operatorname{Re}\lambda - \mu} \\
&\leq C_{\lambda,\mu,\nu}\, K_{\operatorname{Re}\nu + 2\operatorname{Re}\lambda - \mu},
\end{aligned}
$$

whence

$$
|K_\nu * (-\Delta_\mathfrak{v})^\mu u_\lambda| \leq C'_{\lambda,\mu,\nu}\, u_{\operatorname{Re}\lambda + \operatorname{Re}\nu/2 - \mu/2}.
$$

Using this reasoning, it follows that $(-\Delta_\mathfrak{v})^{\mu-\nu} u_\lambda$ is a well defined function, belonging to $L^2(N)$ if $2\operatorname{Re}\lambda + \operatorname{Re}\nu - \mu < r$. We conclude that $(-\Delta_\mathfrak{v})^\lambda u_\lambda$ may be defined by the above process for λ in S, is analytic in λ, and is in $L^2(N)$, as required. $\qquad\square$

From this lemma, it is possible to combine formulae (5.10) and (5.13), and continue analytically into S, to obtain the following result.

Corollary 5.4.2. *For all λ in S,*

$$
\varphi_\lambda = \frac{2^{q-1}\,\Gamma(r+r_0)}{\pi^{r+r_0}} \, \big\langle \pi_0(\cdot)(-\Delta_\mathfrak{v})^\lambda u_\lambda, (-\Delta_\mathfrak{v})^{-\bar\lambda} u_{-\bar\lambda} \big\rangle.
$$

Further progress will depend on being able to analyse $(-\Delta_\mathfrak{v})^\lambda u_\lambda$ on N. We will not be able to give a formula for $\big((-\Delta_\mathfrak{v})^\lambda u_\lambda\big)\hat{}\,(h, \eta)$ in terms of known functions, but an auxiliary function, defined by an integral, will be needed. We define $M: \mathbb{R}^+ \times \mathbb{C} \times \mathbb{C} \to \mathbb{C}$ by the formula

$$
M(\eta, a, b) = \frac{(2\eta)^a}{\Gamma(a)} \int_{\mathbb{R}+} \frac{t^{a-1}}{(t+1)^b}\, e^{-2\eta t}\, dt.
$$

A priori, $M(\eta, a, b)$ is only defined if $\operatorname{Re} a > 0$, since otherwise the integral does not converge. However, by writing the integral as an integral over $]0, 1/2[$ and another over $]1/2, \infty[$, and expanding $t \mapsto (t+1)^{-b} e^{-2\eta t}$ as a Maclaurin series, it may be shown that the expression defined when $\operatorname{Re} a > 0$ continues analytically to an entire function. Further, we have the functional equation

$$
M(\eta, a, b) = M(\eta, b, a) \qquad \forall \eta \in \mathbb{R}^+ \quad \forall a, b \in \mathbb{C}
$$

(see [CH89, Prop. 3.6]). Indeed, for η in $\mathbb{R}^+$, and for functions $f: \mathbb{R}^+ \to \mathbb{C}$ and $g: \mathbb{R}^+ \to \mathbb{C}$ such that $(s, t) \mapsto f(s)\, g(t)\, e^{-2\eta st}$ is integrable on $\mathbb{R}^+ \times \mathbb{R}^+$, with

Laplace transforms $\tilde{f}$ and $\tilde{g}$, we see that

$$\int_{\mathbb{R}^+} \tilde{f}(2\eta t)\, g(t)\, dt = \int_{\mathbb{R}^+} \int_{\mathbb{R}^+} f(s)\, e^{-2\eta st}\, g(t)\, ds\, dt$$

$$= \int_{\mathbb{R}^+} \int_{\mathbb{R}^+} f(s)\, e^{-2\eta st}\, g(t)\, dt\, ds$$

$$= \int_{\mathbb{R}^+} f(s)\, \tilde{g}(2\eta s)\, ds.$$

Applying this formula with f and g given by

$$f(s) = \frac{(2\eta)^a}{\Gamma(a)}\, s^{a-1}\, e^{-2\eta s} \qquad \forall s \in \mathbb{R}^+$$

$$g(t) = \frac{(2\eta)^b}{\Gamma(b)}\, t^{b-1}\, e^{-2\eta t} \qquad \forall t \in \mathbb{R}^+$$

establishes the functional equation when $\operatorname{Re} a > 0$ and $\operatorname{Re} b > 0$; analytic continuation gives it in full generality.

Lemma 5.4.3. *Let u_λ be the function defined in* (5.12). *If* $\operatorname{Re}\lambda < r_0$, *then $\hat{u}_\lambda(h,\eta)$ is equal to*

$$\frac{2^{r_0-h+\lambda}\, \eta^{-r_0-h-\lambda}\, \pi^{r+r_0}\, \Gamma(r_0+h-\lambda)\, e^{-\eta}}{\Gamma(r-\lambda)\, \Gamma(r_0-\lambda)}\, M(\eta, r_0+h-\lambda, r_0+h-\lambda).$$

Proof. The proof involves several steps.

First, we claim that, if $\operatorname{Re}\lambda < r_0$ and $c \in \mathbb{R}^+$, then

$$\int_{\mathbb{R}^q} \left(c^2 + |z|^2\right)^{\lambda-r} j_{(q)}(\eta z)\, dz$$

$$= \frac{\pi^{r-r_0}\, \Gamma(r_0-\lambda)}{\Gamma(r-\lambda)} \int_{\mathbb{R}} \left(c^2 + |s|^2\right)^{\lambda-r_0} e^{-i\eta s}\, ds. \qquad (5.14)$$

If $q = 1$, this formula is trivially true. Otherwise, following [CH89, Lem. 3.4], we take a unit vector E in $\mathbb{R}^q$; then the left-hand side of (5.14) is equal to

$$\int_{\mathbb{R}^q} \left(c^2 + |z|^2\right)^{\lambda-r} e^{-i\eta z \cdot E}\, dz$$

$$= \int_{\mathbb{R}} \int_{E^\perp} \left(c^2 + s^2 + |w|^2\right)^{\lambda-r} e^{-i\eta s}\, dw\, ds$$

$$= \omega_{q-2} \int_{\mathbb{R}} \int_{\mathbb{R}^+} \left(c^2 + s^2 + t^2\right)^{\lambda-r} e^{-i\eta s}\, t^{q-2}\, dt\, ds$$

$$= \omega_{q-2} \int_{\mathbb{R}} \left(c^2 + s^2\right)^{\lambda-r} e^{-i\eta s}\, \frac{\Gamma(r_0 - \lambda)\,\Gamma(r - r_0)}{2\,\Gamma(r - \lambda)}\, ds$$

$$= \frac{\pi^{r-r_0}\,\Gamma(r_0 - \lambda)}{\Gamma(r - \lambda)} \int_{\mathbb{R}} \left(c^2 + s^2\right)^{\lambda - r_0} e^{-i\eta s}\, ds,$$

as claimed.

Our next step is to show that, if $c \in \mathbb{R}^+$ and $\operatorname{Re}\lambda < r_0$, then

$$\int_{\mathbb{R}} (c^2 + s^2)^{\lambda - r_0}\, e^{-i\eta s}\, ds$$

$$= \frac{2\,\pi\,\eta^{2r_0 - 2\lambda - 1}}{\Gamma(r_0 - \lambda)^2} \int_{\mathbb{R}+} t^{r_0 - \lambda - 1}\, (t+1)^{r_0 - \lambda - 1}\, e^{-c\eta(2t+1)}\, dt \qquad (5.15)$$

(see [CH89, Thm 3.5]). To see this, recall first that for functions f and g in $L^2(\mathbb{R})$, the Plancherel theorem states that

$$\int_{\mathbb{R}} \hat{f}(s)\, \overline{\hat{g}}(s)\, ds = 2\,\pi \int_{\mathbb{R}} f(t)\, \overline{g}(t)\, dt.$$

We apply this with f and g given by

$$f(t) = \frac{(t_+)^{\gamma-1}\, e^{-ct}}{\Gamma(\gamma)} \quad \text{and} \quad g(t) = \frac{((t+\eta)_+)^{\gamma-1}\, e^{-c(t+\eta)}}{\Gamma(\gamma)},$$

where t_+ is equal to t when t is positive and zero otherwise, and $\gamma \in\,]1, \infty[$, and deduce that

$$\int_{\mathbb{R}} (c+is)^{-\gamma} \left((c+is)^{-\gamma} e^{i\eta s}\right)^{-}\, ds$$

$$= \frac{2\,\pi}{\Gamma(\gamma)^2} \int_{\mathbb{R}} (t_+)^{\gamma-1} ((t+\eta)_+)^{\gamma-1}\, e^{-c(2t+\eta)}\, dt,$$

whence

$$\int_{\mathbb{R}} (c^2 + s^2)^{-\gamma}\, e^{-i\eta s}\, ds$$

$$= \frac{2\,\pi\,\eta^{2\gamma-1}}{\Gamma(\gamma)^2} \int_{\mathbb{R}+} t^{\gamma-1}\, (t+1)^{\gamma-1}\, e^{-c\eta(2t+1)}\, dt.$$

Analytic continuation of γ into $\{\gamma \in \mathbb{C} : \operatorname{Re}\gamma > 0\}$ proves the desired result.

Now we can complete the proof. From formulae (5.1), (5.14) and (5.15), and some changes of variables, we see that

$$
\int_{\mathbb{R}^p} \int_{\mathbb{R}^q} \left(\left(1 + \frac{|x|^2}{4} \right)^2 + |z|^2 \right)^{\lambda - r} \psi_{h,\eta}(x, z) \, dz \, dx
$$

$$
= \int_{\mathbb{R}^p} \int_{\mathbb{R}^q} \left(\left(1 + \frac{|x|^2}{4} \right)^2 + |z|^2 \right)^{\lambda - r} \binom{h + \alpha}{h}^{-1}
$$
$$
\exp\left(-\frac{\eta |x|^2}{4} \right) L_h^{(\alpha)} \left(\frac{\eta |x|^2}{2} \right) j_{(q)}(\eta z) \, dz \, dx
$$

$$
= \binom{h + \alpha}{h}^{-1} \frac{\pi^{r - r_0} \Gamma(r_0 - \lambda)}{\Gamma(r - \lambda)} \int_{\mathbb{R}^p} \int_{\mathbb{R}} \left(\left(1 + \frac{|x|^2}{4} \right)^2 + s^2 \right)^{\lambda - r_0}
$$
$$
\exp\left(-\frac{\eta |x|^2}{4} \right) L_h^{(\alpha)} \left(\frac{\eta |x|^2}{2} \right) \exp(-i\eta s) \, ds \, dx
$$

$$
= \binom{h + \alpha}{h}^{-1} \frac{\pi^{r - r_0} \Gamma(r_0 - \lambda)}{\Gamma(r - \lambda)} \int_{\mathbb{R}^p} \frac{2\pi \eta^{2 r_0 - 2\lambda - 1}}{\Gamma(r_0 - \lambda)^2} \int_{\mathbb{R}^+} t^{r_0 - \lambda - 1} (t + 1)^{r_0 - \lambda - 1}
$$
$$
\exp\left(-\eta \left(1 + \frac{|x|^2}{4} \right)(2t + 1) - \frac{\eta |x|^2}{4} \right) L_h^{(\alpha)} \left(\frac{\eta |x|^2}{2} \right) \, dt \, dx
$$

$$
= \binom{h + \alpha}{h}^{-1} \frac{2 \pi^{r - r_0 + 1} \eta^{2 r_0 - 2\lambda - 1}}{\Gamma(r - \lambda) \Gamma(r_0 - \lambda)} \omega_{p-1} \int_{\mathbb{R}^+} \int_{\mathbb{R}^+} t^{r_0 - \lambda - 1} (t + 1)^{r_0 - \lambda - 1}
$$
$$
\exp\left(-\eta \left(1 + \frac{s^2}{4} \right)(2t + 1) - \frac{\eta s^2}{4} \right) L_h^{(\alpha)} \left(\frac{\eta s^2}{2} \right) s^{p-1} \, ds \, dt
$$

$$
= \binom{h + \alpha}{h}^{-1} \frac{2^{2 r_0} \pi^{r + r_0} \eta^{-2\lambda}}{\Gamma(r - \lambda) \Gamma(r_0 - \lambda) \Gamma(\alpha + 1)} \int_{\mathbb{R}^+} \int_{\mathbb{R}^+} t^{r_0 - \lambda - 1} (t + 1)^{r_0 - \lambda - 1}
$$
$$
\exp(-\eta(2t + 1) - v(t + 1)) L_h^{(\alpha)}(v) v^\alpha \, dv \, dt
$$

$$
= \binom{h + \alpha}{h}^{-1} \frac{2^{2 r_0} \pi^{r + r_0} \eta^{-2\lambda}}{\Gamma(r - \lambda) \Gamma(r_0 - \lambda) \Gamma(\alpha + 1)} \int_{\mathbb{R}^+} \int_{\mathbb{R}^+} t^{r_0 - \lambda - 1} (t + 1)^{r_0 - \lambda - 1}
$$
$$
\exp(-\eta(2t + 1)) \frac{\Gamma(h + \alpha + 1) t^h}{h! (t + 1)^{h + \alpha + 1}} \, dt
$$

by formula (5.18). Thus

$$
\hat{u}_\lambda(h, \eta) = \frac{2^{2 r_0} \pi^{r + r_0} \eta^{-2\lambda} e^{-\eta}}{\Gamma(r - \lambda) \Gamma(r_0 - \lambda)} \int_{\mathbb{R}^+} \frac{t^{r_0 + h - \lambda - 1}}{(t + 1)^{r_0 + h + \lambda}} e^{-2\eta t} \, dt
$$

$$
= \frac{2^{2 r_0} \pi^{r + r_0} \eta^{-2\lambda} e^{-\eta} \Gamma(r_0 + h - \lambda)}{\Gamma(r - \lambda) \Gamma(r_0 - \lambda) (2\eta)^{r_0 + h - \lambda}} M(\eta, r_0 + h - \lambda, r_0 + h + \lambda),
$$

and the result required holds. $\square$

Corollary 5.4.4. *If $\lambda \in S$, then $\left((-\Delta_v)^\lambda u_\lambda\right)^{\hat{}}(h,\eta)$ is equal to*

$$\frac{2^{2r_0+\lambda}\pi^{r+r_0}(\alpha+2h+1)^\lambda \Gamma(r_0+h-\lambda)}{(2\eta)^{r_0+h}e^\eta \Gamma(r-\lambda)\Gamma(r_0-\lambda)} M(\eta, r_0+h-\lambda, r_0+h+\lambda).$$

Proof. This follows from the definition of $(-\Delta_v)^\lambda$ and the previous lemma when $\text{Re}\,\lambda < r_0$. It follows for general λ in S by analytic continuation. $\square$

Let $v_{h,\lambda}$ be the $L^2(N)$-function such that $(v_{h,\lambda})^{\hat{}}(h',\eta)$ is equal to 0 if $h \neq h'$ and

$$(v_{h,\lambda})^{\hat{}}(h',\eta) = \frac{M(\eta, r_0+h-\lambda, r_0+h+\lambda)}{(2\eta)^{r_0+h}e^\eta}.$$

Then we have the following corollary.

Corollary 5.4.5. *When $\lambda \in (-r, r)$, φ_λ is equal to*

$$(4\pi)^{r+r_0}\Gamma(r+r_0)\sum_{h\in\mathbb{N}}\frac{\Gamma(r_0+h-\lambda)\Gamma(r_0+h+\lambda)}{\Gamma(r-\lambda)\Gamma(r_0-\lambda)\Gamma(r+\lambda)\Gamma(r_0+\lambda)}\langle \pi_0(\cdot)v_{h,\lambda}, v_{h,\lambda}\rangle.$$

This sum converges absolutely. Further, φ_λ is positive definite if and only if $\lambda \in [-r_0, r_0]$.

Proof. The Fourier transform calculation of the previous corollary shows that

$$P_h\left((-\Delta_v)^\lambda u_\lambda\right) = \frac{2^{2r_0+\lambda}\pi^{r+r_0}(\alpha+2h+1)^\lambda \Gamma(r_0+h-\lambda)}{\Gamma(r-\lambda)\Gamma(r_0-\lambda)}v_{h,\lambda}.$$

From Corollary 5.4.2 and formula (5.11), we deduce that φ_λ is equal to

$$\frac{2^{q-1}\Gamma(r+r_0)}{\pi^{r+r_0}}\langle \pi_0(\cdot)(-\Delta_v)^\lambda u_\lambda, (-\Delta_v)^{-\lambda}u_{-\lambda}\rangle$$

$$= \frac{2^{q-1}\Gamma(r+r_0)}{\pi^{r+r_0}}\sum_{h\in\mathbb{N}}\langle \pi_0(\cdot)P_h(-\Delta_v)^{-\lambda}u_{-\lambda}, P_h(-\Delta_v)^{-\lambda}u_{-\lambda}\rangle$$

$$= \frac{2^{q-1}\Gamma(r+r_0)}{\pi^{r+r_0}}\sum_{h\in\mathbb{N}}\frac{2^{4r_0}\pi^{2(r+r_0)}\Gamma(r_0+h-\lambda)\Gamma(r_0+h+\lambda)}{\Gamma(r-\lambda)\Gamma(r_0-\lambda)\Gamma(r+\lambda)\Gamma(r_0+\lambda)}\langle \pi_0(\cdot)v_{h,\eta}, v_{h,\eta}\rangle$$

$$= (4\pi)^{r+r_0}\Gamma(r+r_0)\sum_{h\in\mathbb{N}}\frac{\Gamma(r_0+h-\lambda)\Gamma(r_0+h+\lambda)}{\Gamma(r-\lambda)\Gamma(r_0-\lambda)\Gamma(r+\lambda)\Gamma(r_0+\lambda)}\langle \pi_0(\cdot)v_{h,\eta}, v_{h,\eta}\rangle,$$

as enunciated. The convergence of this series is a particular case of (5.9). Finally, if each of the quotients

$$\frac{\Gamma(r_0+h-\lambda)\Gamma(r_0+h+\lambda)}{\Gamma(r-\lambda)\Gamma(r_0-\lambda)\Gamma(r+\lambda)\Gamma(r_0+\lambda)}$$

(where $h \in \mathbb{N}$) is nonnegative, then φ_λ is positive definite. Conversely, by formula (5.8),

$$\langle \pi_0(\cdot) v_{h,\lambda}, v_{h,\lambda} \rangle \big|_N = \overline{v}_{h,\lambda} * (v_{h,\lambda})^{\widehat{}} = \overline{v}_{h,\lambda} * v_{h,\lambda},$$

whence

$$\left(\langle \pi_0(\cdot) v_{h,\lambda}, v_{h,\lambda} \rangle \big|_N \right)^{\widehat{}} = (\overline{v}_{h,\lambda})^{\widehat{}} (v_{h,\lambda})^{\widehat{}} = \left| (v_{h,\lambda})^{\widehat{}} \right|^2 .$$

Further, $\eta \mapsto (v_{h,\lambda})^{\widehat{}}(h, \eta)$ does not vanish identically, and it follows that $\left(\langle \pi_0(\cdot) v_{h,\lambda}, v_{h,\lambda} \rangle \big|_N \right)^{\widehat{}}$ is nonnegative only if each of the quotients

$$\frac{\Gamma(r_0 + h - \lambda)\,\Gamma(r_0 + h + \lambda)}{\Gamma(r - \lambda)\,\Gamma(r_0 - \lambda)\,\Gamma(r + \lambda)\,\Gamma(r_0 + \lambda)}$$

is nonnegative. For λ in $]0, r[$, the only factors of the quotient where sign changes might occur are $\Gamma(r_0 + h - \lambda)$ and $\Gamma(r_0 - \lambda)$. Since

$$\frac{\Gamma(r_0 + h - \lambda)}{\Gamma(r_0 - \lambda)} = (r_0 + h - 1 - \lambda) \dots (r_0 - \lambda),$$

sign changes occur when $\lambda = r_0, r_0 + 1, r_0 + 2, \dots$.

   When $h = 0$, there is no sign change, while if $h = 1$ there is one sign change, when $\lambda = r_0$. Thus when $\lambda \in (r_0, r)$, the coefficient of $\langle \pi_0(\cdot) v_{0,\eta}, v_{0,\eta} \rangle$ is positive, while that of $\langle \pi_0(\cdot) v_{1,\lambda}, v_{1,\lambda} \rangle$ is negative, so φ_λ is not positive definite. $\square$

Remarks. These results show that the groups $\mathrm{Sp}(n, 1)$ and $F_{4,-20}$ do not have the Haagerup property. Indeed, if there were a sequence of positive definite functions vanishing at infinity which tended to 1, then there would be a net of positive definite K-invariant functions. The restriction of these to the Iwasawa NA subgroup would contradict our result here. Combined with the Howe–Moore property, this implies that these groups have property (T).

   We should also remark that it is reasonably simple to extend the results of [CH89] to define a "radial Haagerup constant" $\Lambda_{NA}^{\mathrm{rad}}$ for the harmonic NA groups, and compute this. Its value is

$$\frac{\pi^{1/2}\,\Gamma((p+q)/2)}{\Gamma((q/2)\,\Gamma((p+1)/2)}$$

(see [CH89, Sec. 5]).

5.5   Appendix on special functions

We use two sorts of special functions in our analysis on N, and it seems useful to summarize the essential properties of these functions and outline their proofs.

We define the radial functions $j_{(n)} \colon \mathbb{R}^n \to \mathbb{C}$ by the formula

$$j_{(n)}(x) = \int_{SO(n)} \exp(i\sigma x \cdot E)\, d\sigma$$

where E is a fixed but arbitrary unit vector in $\mathbb{R}^n$ and the integration is relative to normalized Haar measure on $SO(n)$. If $f \colon \mathbb{R}^n \to \mathbb{C}$ is radial, then

$$\int_{\mathbb{R}^{(n)}} j_{(n)}(\eta x)\, f(x)\, dx = \int_{SO(n)} \int_{\mathbb{R}^n} e^{-i\sigma \eta x \cdot E} f(x)\, dx\, d\sigma$$

$$= \int_{SO(n)} \int_{\mathbb{R}^n} e^{-iy \cdot \eta E} f(y)\, dy\, d\sigma$$

$$= \int_{\mathbb{R}^n} e^{-iy \cdot \eta E} f(y)\, dy$$

$$= \hat{f}(\eta E).$$

Note that there is an even function $J_{(n)} \colon \mathbb{R} \to \mathbb{C}$ such that $j_{(n)}(x) = J_{(n)}(|x|)$. Since $\Delta(f \circ \sigma) = (\Delta f) \circ \sigma$ for any function f on $\mathbb{R}^n$ and any σ in $SO(q)$, we may conclude that $\Delta j_{(n)} = j_{(n)}$; writing the Laplacian in polar coordinates shows that

$$\left(\frac{d^2}{dt^2} + \frac{n-1}{t} \frac{d}{dt} \right) J_{(n)}(t) = -J_{(n)}(t) \qquad \forall t \in \mathbb{R}^+.$$

Since $J_{(n)}$ is clearly analytic and $J_{(n)}(0) = 1$, it follows that

$$J_{(n)}(t) = \sum_{l \in \mathbb{N}} \frac{\Gamma(n/2)}{l!\,\Gamma(l + n/2)} \left(\frac{-t^2}{4} \right)^l.$$

This is a modification of the standard Bessel function $J_{n/2-1}$:

$$J_{(n)}(t) = \left(\frac{2}{t} \right)^{n/2-1} \Gamma\left(\frac{n}{2} \right) J_{n/2-1}(t).$$

We define the (generalized) Laguerre polynomial $L_n^{(\alpha)}$ by Rodrigues' formula:

$$L_n^{(\alpha)}(x) = \frac{e^x x^{-\alpha}}{n!} \frac{d^n}{dx^n} \left(e^{-x} x^{n+\alpha} \right)$$

$$= \frac{e^x x^{-\alpha}}{n!} \sum_{l=0}^{n} \binom{n}{l} (-1)^l e^{-x} \frac{\Gamma(n + \alpha + 1)}{\Gamma(l + \alpha + 1)} x^{\alpha+l}$$

$$= \sum_{l=0}^{n} \frac{(-1)^l \Gamma(n + \alpha + 1)}{l!\,(n-l)!\,\Gamma(l + \alpha + 1)} e^{-x} x^l. \qquad (5.16)$$

In particular, the constant term of $L_n^{(\alpha)}$ is $\Gamma(n + \alpha + 1)/(n!\,\Gamma(\alpha + 1))$.

From formula (5.16), it is a simple but tedious verification that, if

$$y(x) = e^{-x/2} \, x^{(\alpha+1)/2} \, L_n^{(\alpha)}(x),$$

then

$$\frac{d^2 y(x)}{dx^2} + \left(\frac{2n + \alpha + 1}{2x} + \frac{1 - \alpha^2}{4x^2} - \frac{1}{4} \right) y(x) = 0.$$

Other differential equations may be checked similarly.

Rodrigues' formula and integration by parts lead to some integral formulae: for instance, for k in $\mathbb{N}$,

$$
\begin{aligned}
\int_0^\infty L_n^{(\alpha)}(x) \, x^k \, e^{-x} x^\alpha \, dx &= \frac{1}{n!} \int_0^\infty \frac{d^n}{dx^n} \left(e^{-x} x^{n+\alpha} \right) x^k \, dx \\
&= \frac{(-1)^n}{n!} \int_0^\infty e^{-x} x^{n+\alpha} \frac{d^n x^k}{dx^n} \, dx \\
&= \frac{(-1)^n \, \Gamma(k + \alpha + 1) \, \Gamma(k + 1)}{n! \, \Gamma(k + 1 - n)}
\end{aligned}
$$

(this is to be interpreted as 0 if $k < n$). It now follows from formula (5.16) that

$$\int_0^\infty L_m^{(\alpha)}(x) \, L_n^{(\alpha)}(x) \, e^{-x} x^\alpha \, dx = \begin{cases} 0 & \text{if } m < n \\ \Gamma(n + \alpha + 1)/n! & \text{if } m = n. \end{cases} \quad (5.17)$$

In conclusion, we see similarly that, if $k \in \mathbb{R}^+$, then

$$
\begin{aligned}
\int_0^\infty L_n^{(\alpha)}(x) \, e^{-kx} \, e^{-x} x^\alpha \, dx &= \frac{1}{n!} \int_0^\infty \frac{d^n}{dx^n} \left(e^{-x} x^{n+\alpha} \right) e^{-kx} \, dx \\
&= \frac{k^n}{n!} \int_0^\infty e^{-x} x^{n+\alpha} e^{-kx} \, dx \\
&= \frac{\Gamma(n + \alpha + 1) \, k^n}{n! \, (k + 1)^{n+\alpha+1}}. \quad (5.18)
\end{aligned}
$$

Chapter 6

Discrete Groups
by Paul Jolissaint, Pierre Julg and Alain Valette

The main theme of this chapter is the behaviour of the Haagerup property under various group constructions[1]. In the first section, we deal with inductive limits and group extensions. In the second, we prove that if a group Γ acts on a tree, with finite edge stabilizers, and with vertex stabilizers having the Haagerup property, then Γ itself has the Haagerup property. In particular the Haagerup property is preserved under free products, or amalgamated products over finite groups (see also [Jol00] for a different proof of this fact). The third section has a somewhat different flavour: we give a sufficient condition for a finitely presented group to have the Haagerup property and simultaneously be of cohomological dimension at most 2 (in particular the group must be torsion-free).

6.1 Some hereditary results

Proposition 6.1.1. *Assume that the locally compact group G is the increasing union of a sequence $(G_n)_{n \geq 1}$ of open subgroups. If all G_n have the Haagerup property, then so does G.*

Proof. If ϕ is a positive definite function on some G_n, we denote by $\tilde{\phi}$ the positive definite function on G equal to ϕ on G_n and to 0 outside G_n. For each

---

[1]This chapter was circulated in preprint form at the beginning of 1998, under the name "Nouveaux exemples de groupes avec la propriété de Haagerup". We thank S.R. Gal and T. Januszkiewicz for pointing out a mistake in the preliminary version.

n, let $(\phi_{n,k})_{k\geq 1}$ be a sequence of normalized positive definite functions on G_n, vanishing at infinity and converging to 1 uniformly on compact subsets of G_n. Allowing n to vary, we consider the family $(\tilde{\phi}_{n,k})_{n,k\geq 1}$ of normalized, positive definite functions on G, vanishing at infinity. By a diagonalization process, we may extract a sequence converging to 1 uniformly on compact subsets of G. $\square$

Example 6.1.2 (Adèle groups). First, let K be a global field, and denote the ring of adèles of K by $\mathbb{A}_K$. Then the groups $\mathrm{SL}_2(K)$ and $\mathrm{SL}_2(\mathbb{A}_K)$ have the Haagerup property.

To see this, note that $\mathrm{SL}_2(K)$ is a closed subgroup of $\mathrm{SL}_2(\mathbb{A}_K)$, so that it is enough to show that $G = \mathrm{SL}_2(\mathbb{A}_K)$ has the Haagerup property. For that, denote the set of all places of K by P and the (finite) subset of P of archimedean places by P_∞; denote the completion of K at v by K_v for $v \in P$ and the ring of integers of K_v by $\mathcal{O}_v$ for $v \in P \setminus P_\infty$. Recall that G is the set of sequences $(g_v)_{v\in P}$ such that $g_v \in \mathrm{SL}_2(K_v)$ for all v, and $g_v \in \mathrm{SL}_2(\mathcal{O}_v)$ for all but a finite number of v in $P \setminus P_\infty$. So G is the inductive limit, as S varies over the finite subsets of P containing P_∞, of the open subgroups

$$\prod_{v\in S} \mathrm{SL}_2(K_v) \times \prod_{v\in P\setminus S} \mathrm{SL}_2(\mathcal{O}_v); \tag{6.1}$$

each of these has the Haagerup property, by Examples 1.2.2 and 1.2.3, so we may appeal to Proposition 6.1.1.

It seems worthwhile to give an explicit, proper, conditionally negative definite function on G. To do this, we need some more notation. For $v \in P\setminus P_\infty$, let π_v be a uniformizer in $\mathcal{O}_v$; moreover, let X_v be the tree associated with $\mathrm{SL}_2(K_v)$ (see [Ser77]), d_v be the distance function on X_v, and x_v be the vertex of X_v associated with the maximal compact subgroup $\mathrm{SL}_2(\mathcal{O}_v)$. We begin by defining, for each place $v \in P$, a conditionally negative definite function ψ_v on G; if v is archimedean, we define ψ_v as the composition of the canonical projection $G \to \mathrm{SL}_2(K_v)$ with a continuous, conditionally negative definite function which is proper on $\mathrm{SL}_2(K_v)$ (such as the distance to the origin in real hyperbolic space of dimension 2 if $K_v = \mathbb{R}$ or of dimension 3 if $K_v = \mathbb{C}$, see Example 1.2.2); if v is nonarchimedean, as in Example 1.2.3, we set

$$\psi_v(g) = d_v(g_v x_v, x_v)$$

(where $g = (g_w)_{w\in P} \in G$; note that ψ_v is integer-valued). Finally, we define, for $g \in G$

$$\Psi(g) = \sum_{v\in P_\infty} \psi_v(g) + \sum_{v\in P\setminus P_\infty} |\pi_v|_v^{-1}\, \psi_v(g). \tag{6.2}$$

By construction, the right-hand side of (6.2) is a finite sum, so that Ψ is a continuous, conditionally negative definite function on G. To check that Ψ is proper, we fix $T > 0$. There exist finitely many nonarchimedean places v such that $|\pi_v|_v^{-1} \le T$. Denote by S the set of all these places and the infinite places. If $g \in G$ is such that $\Psi(g) \le T$, then in fact

$$g \in \prod_{v \in S} \mathrm{SL}_2(K_v) \times \prod_{v \in P \backslash S} \mathrm{SL}_2(\mathcal{O}_v),$$

and the restriction of Ψ to this direct product is clearly proper.

The same argument shows that the groups $\mathrm{PGL}_2(K)$ and $\mathrm{PGL}_2(\mathbb{A}_K)$ also have the Haagerup property.

Here is a second example. Again, let K be a global field, this time not of characteristic 2, and let H be a quaternion algebra over K. Denote by H^1 the group of quaternions with reduced norm 1, viewed as an algebraic group defined over K. The groups $H^1(K)$ of points over K, and $H^1(\mathbb{A}_K)$ of adelic points, have the Haagerup property. Again, since $H^1(K)$ is discrete in $H^1(\mathbb{A}_K)$, it is enough to prove it for $H^1(\mathbb{A}_K)$. The reasoning is the same as in the preceding example, by noticing that $H^1(K_v)$ is compact if H ramifies at v (that is, if $H \otimes_K K_v$ is a division algebra); otherwise $H^1(K_v) \simeq \mathrm{SL}_2(K_v)$; see [Vig80, p. 81], for all this.

The next lemma is considered "obvious" in [Gro88, 4.5.C]. It says that inducing a C_0-representation yields a C_0-representation; this answers positively a question of L. Baggett and K. Taylor [BT79], who proved the lemma under the extra assumption that the C_0-representation is induced either from an open subgroup, or from a central subgroup ([BT79, Lem. A and B]). Another proof of Lemma 6.1.3 may be found in [Jol00], somewhat hidden in the proof of Theorem 2.7. The proof given here was almost entirely supplied by B. Bekka.

Lemma 6.1.3. *Let H be a closed subgroup of the locally compact group G, and let π be a C_0-representation of H. Then the induced representation $\mathrm{Ind}_H^G \pi$ is a C_0-representation of G.*

Proof. As is common with induced representations, we need much notation for the proof. We shall roughly follow the presentation of G. Warner ([War72, 5.1.1 and Appendix A.1]).

Let $\rho \colon G \to \mathbb{R}^+$ be a Bruhat function, that is, a continuous positive function such that $\rho(1) = 1$ and

$$\rho(xh) = \frac{\Delta_H(h)}{\Delta_G(h)} \rho(x) \qquad \forall x \in G \quad \forall h \in H; \tag{6.3}$$

here Δ_G and Δ_H are the modular homomorphisms of G and H. Such a Bruhat function defines a quasi-invariant measure $d\dot{x}$ on G/H by the formula

$$\int_{G/H} \phi^H(\dot{x})\, d\dot{x} = \int_G \phi(x)\, \rho(x)\, dx, \qquad (6.4)$$

for all continuous functions ϕ with compact support on G, where

$$\phi^H(x) = \int_H \phi(xh)\, dh \qquad \forall x \in G.$$

Of course, we abuse notation in the standard way in (6.4), by viewing ϕ^H as a function on G/H.

We write σ for $\mathrm{Ind}_H^G \pi$, and proceed to describe σ carefully. The underlying Hilbert space $\mathcal{H}_\sigma$ is the space of measurable functions $\eta\colon G \to \mathcal{H}_\pi$ satisfying

$$\eta(xh) = \pi(h^{-1})\eta(x) \qquad \forall h \in H,$$

for almost every $x \in G$, and such that

$$\int_{G/H} \|\eta(\dot{x})\|^2\, d\dot{x} < \infty.$$

The group G acts unitarily on $\mathcal{H}_\sigma$ by

$$(\sigma(g)\eta)(x) = \eta(g^{-1}x)\sqrt{\frac{\rho(g^{-1}x)}{\rho(x)}}$$

for all g and $x \in G$ and all $\eta \in \mathcal{H}_\sigma$.

For ξ in $\mathcal{H}_\pi$ and a continuous function ϕ with compact support on G, we define

$$I_{\phi,\xi}(x) = \int_H \phi(xh)\,\pi(h)\xi\, dh \qquad \forall x \in G.$$

It is well known and easy to prove that $I_{\phi,\xi}$ is in $\mathcal{H}_\sigma$, and that as ϕ and ξ vary, the functions $I_{\phi,\xi}$ span a dense subspace of $\mathcal{H}_\sigma$.

Therefore, to prove that σ is a C_0-representation, it is enough to show that the matrix coefficients

$$g \mapsto \langle \sigma(g) I_{\phi,\xi}, I_{\phi,\xi} \rangle$$

vanish at infinity on G.

For that, let C be the support of ϕ. Then

$$\langle \sigma(g) I_{\phi,\xi}, I_{\phi,\xi} \rangle$$

$$= \int_{G/H} \sqrt{\frac{\rho(g^{-1}\dot{x})}{\rho(\dot{x})}} \int_H \int_H \phi(g^{-1}xh)\overline{\phi}(xh') \langle \pi(h)\xi, \pi(h')\xi \rangle \, dh \, dh' \, d\dot{x}$$

$$= \int_{G/H} \sqrt{\frac{\rho(g^{-1}\dot{x})}{\rho(\dot{x})}} \int_H \overline{\phi}(xh') \int_H \phi(g^{-1}xh) \left\langle \xi, \pi((h'^{-1}h)^{-1})\xi \right\rangle \, dh \, dh' \, d\dot{x}$$

$$= \int_H \langle \xi, \pi(h^{-1})\xi \rangle \int_{G/H} \int_H \phi(g^{-1}xh'h)\overline{\phi}(xh')\sqrt{\frac{\rho(g^{-1}xh')}{\rho(xh')}} \, dh' \, d\dot{x} \, dh$$

$$= \int_H \langle \xi, \pi(h^{-1})\xi \rangle \int_G \phi(g^{-1}xh)\overline{\phi}(x)\sqrt{\rho(g^{-1}x)\rho(x)} \, dx \, dh$$

$$= \int_H \langle \xi, \pi(h^{-1})\xi \rangle \int_G \phi(g^{-1}x)\overline{\phi}(xh^{-1})\sqrt{\rho(g^{-1}xh^{-1})\rho(xh^{-1})}\Delta_G(h^{-1}) \, dx \, dh$$

$$= \int_H \langle \xi, \pi(h^{-1})\xi \rangle \int_G \phi(g^{-1}x)\overline{\phi}(xh^{-1})\sqrt{\rho(g^{-1}x)\rho(x)} \, dx \, \Delta_H(h^{-1}) \, dh$$

$$= \int_H \langle \xi, \pi(h)\xi \rangle \int_G \phi(g^{-1}x)\overline{\phi}(xh)\sqrt{\rho(g^{-1}x)\rho(x)} \, dx \, dh$$

$$= \int_H \langle \xi, \pi(h)\xi \rangle \int_{gC\cap Ch^{-1}} \phi(g^{-1}x)\overline{\phi}(xh)\sqrt{\rho(g^{-1}x)\rho(x)} \, dx \, dh, \qquad (6.5)$$

by definition, a little rewriting, the change of variables h to $h'h$, Fubini's Theorem and (6.4), (6.3), (6.4) again, the change of variables h to h^{-1}, and finally the support properties of ϕ.

Fix a small positive ϵ, and let K be a compact subset of H such that $|\langle \xi, \pi(h)\xi \rangle| < \epsilon$ when $h \in H \setminus K$. Write L for the compact subset $CK^{-1}C^{-1}$ of G. If $g \in G \setminus L$, then $gC \cap CK^{-1} = \emptyset$, so from (6.5),

$$|\langle \sigma(g) I_{\phi,\xi}, I_{\phi,\xi} \rangle| \le \epsilon \int_{H\setminus K} \int_G |\phi(g^{-1}x)\phi(xh)| \sqrt{\rho(g^{-1}x)\rho(x)} \, dx \, dh. \qquad (6.6)$$

It remains to see that the double integral on the right-hand side of (6.6) may be bounded above by some constant, independent on $g \in G$. But this double integral is less than

$$\int_H \int_G |\phi(g^{-1}x)\phi(xh)| \sqrt{\rho(g^{-1}x)\rho(x)} \, dx \, dh$$

$$= \int_G |\phi(g^{-1}x)| \, |\phi|^H(x)\sqrt{\frac{\rho(g^{-1}x)}{\rho(x)}} \, \rho(x) \, dx$$

$$= \int_{G/H} |\phi|^H(\dot{x}) \, |\phi|^H(g^{-1}\dot{x})\sqrt{\frac{\rho(g^{-1}\dot{x})}{\rho(\dot{x})}} \, d\dot{x}$$

$$\leq \left(\int_{G/H} (|\phi|^H (\dot{x}))^2 \, d\dot{x} \right)^{1/2} \left(\int_{G/H} (|\phi|^H (g^{-1}\dot{x}))^2 \, \frac{\rho(g^{-1}\dot{x})}{\rho(\dot{x})} \, d\dot{x} \right)^{1/2}$$

$$= \int_{G/H} (|\phi|^H (\dot{x}))^2 \, d\dot{x}$$

by definition, (6.4), the Cauchy–Schwarz inequality and the quasi-invariance of the measure $d\dot{x}$. This concludes the proof. $\square$

We recall a definition from [IN96].

Definition 6.1.4. Let H be a closed subgroup of the locally compact group G. We say that H is **co-Følner** in G if there exists a G-invariant state on $L^\infty(G/H)$ (equivalently, if the homogeneous space G/H is amenable in the sense of Eymard [Eym72]).

The following result was proved in [Jol00, Prop. 2.5], in a different way.

Proposition 6.1.5. *Let H be a closed subgroup of the locally compact group G, which is co-Følner in G. If H has the Haagerup property, then so does G.*

Proof. Let π be a C_0-representation of H, weakly containing the trivial representation 1_H of H. Consider the induced representation $\mathrm{Ind}_H^G \pi$; it is a C_0-representation, by Lemma 6.1.3. Denoting by $\prec$ the relation of weak containment, we have assumed that

$$1_H \prec \pi.$$

By the continuity of induction,

$$\mathrm{Ind}_H^G 1_H \prec \mathrm{Ind}_H^G \pi.$$

But $\mathrm{Ind}_H^G 1_H$ is the quasiregular representation of G on $L^2(G/H)$; since H is co-Følner,

$$1_G \prec \mathrm{Ind}_H^G 1_H,$$

by the equivalent characterizations in [Eym72, pp. 28–29] (see also [IN96, Thm 3.3]). By the transitivity of weak containment,

$$1_G \prec \mathrm{Ind}_H^G \pi,$$

and the proof is complete. $\square$

Example 6.1.6 (Extensions with amenable quotients). Consider a short exact sequence of locally compact groups:

$$1 \to N \to G \to G/N \to 1.$$

If N has the Haagerup property and G/N is amenable, then G has the Haagerup property; this follows immediately from Proposition 6.1.5. This applies, for instance, to the short exact sequence

$$1 \to \mathrm{SL}_2(F) \to \mathrm{GL}_2(F) \overset{\det}{\to} F^\times \to 1,$$

where F is a local field; since $\mathrm{SL}_2(F)$ has the Haagerup property (by Examples 1.2.2 and 1.2.3), so does $\mathrm{GL}_2(F)$.

At this juncture, we recall that, as a consequence of relative property (T), it is not generally true that an extension of a group with the Haagerup property by another one with the Haagerup property has the Haagerup property; for instance, the semidirect products $\mathbb{R}^2 \rtimes \mathrm{SL}_2(\mathbb{R})$ and $\mathbb{Z}^2 \rtimes \mathrm{SL}_2(\mathbb{Z})$ have relative property (T) with respect to the normal subgroup, hence cannot have the Haagerup property.

6.2 Groups acting on trees

Lemma 6.2.1. *Let A be an open subgroup of the locally compact group G.*

(1) *Assume A is compact. If ψ is a continuous, conditionally negative definite function on G, then there exists a continuous conditionally negative definite function ψ' on G such that*

    (a) *ψ' is A-bi-invariant, (that is,*

$$\psi'(aga') = \psi'(g) \qquad \forall g \in G \quad \forall a, a' \in A);$$

    (b) *$\psi'(a) = 0$ for all $a \in A$, and $\psi'(g) \geq 1$ for all $g \in G \setminus A$;*

    (c) *$\psi - \psi'$ is bounded.*

(2) *Assume that A is central and that the pair (G, A) has the generalized Haagerup property. Then, given any compact subset K of G and $0 < \varepsilon < 1$, there exist a compact neighbourhood V of 1 in $\hat{A}$, a family $\{\phi_\chi : \chi \in V\}$ of normalized positive definite functions $\phi_\chi \colon G \to \mathbb{C}$ satisfying the four conditions of Definition 4.2.1 (with respect to K and ε), and a constant $c \in \,]0, 1[$ such that $|\phi_\chi(g)| \leq c$ for every $g \in G \setminus A$.*

Proof. To prove (1), observe that there exists a continuous affine action α of G on a Hilbert space $\mathcal{H}$ such that $\psi(g) = \|\alpha(g)(0)\|^2$ for each $g \in G$. Since A is compact, there exists $\xi \in \mathcal{H}$ such that $\alpha(a)(\xi) = \xi$ for $a \in A$. Denote by δ_A the characteristic function of the base-point in G/A (that is, the left coset of A at the identity). The function ψ' on G defined by

$$\psi'(g) = \|\alpha(g)(\xi)\|^2 + \frac{1}{2}\left\|\lambda_{G/A}(g)\delta_A - \delta_A\right\|^2$$

(where $\lambda_{G/A}$ is the quasiregular representation of G on $L^2(G/A)$), is continuous, conditionally negative definite, and right-A-invariant; since $\psi' = \check{\psi}'$, it is also left-A-invariant. The other conditions are clear, from the construction.

For future reference, we notice that this argument really shows that if A is an open subgroup in G and ψ is a continuous, conditionally negative definite function on G, vanishing on A, then there is a bounded perturbation ψ' of ψ, satisfying these same conditions and the additional condition that $\psi'(g) \geq 1$ for $g \in G \setminus A$.

To prove (2), let $\{\phi'_\chi : \chi \in V\}$ be a family of normalized positive definite functions on G, satisfying the four conditions of Definition 4.2.1 with respect to K and $\varepsilon/2$. By Lemmata 4.2.8 and 4.2.10, the discrete group G/A has the Haagerup property. Taking into account the remark at the end of part (1) of this proof, we find a conditionally negative definite function ψ' on G/A such that $\psi'(\dot{x}) \geq 1$ when $\dot{x} \neq 1$ in G/A. Denote by ψ the lift of ψ' to G, and choose t small enough that $0 \leq 1 - e^{-t\psi(g)} < \varepsilon/2$ for every $g \in K$. Then, setting $\phi_\chi = \phi'_\chi e^{-t\psi}$, we get the desired family with $c = e^{-t} < 1$. $\qquad\square$

Lemma 6.2.2. *Let Γ be a discrete group acting (on the left) on a set X; let H be a group, and let $c \colon X \times \Gamma \to H$ a map verifying the cocycle relation*

$$c(x, \gamma_1\gamma_2) = c(x, \gamma_1)c(\gamma_1^{-1}x, \gamma_2)$$

for all $x \in X$ and $\gamma_1, \gamma_2 \in \Gamma$. Let ψ be a conditionally negative definite function on H, vanishing on a subset A of H. Assume that, for every $\gamma \in \Gamma$, the set $\{x \in X : c(x, \gamma) \notin A\}$ is finite; then the function $\tilde\psi$ on Γ may be defined by

$$\tilde\psi(\gamma) = \sum_{x \in X} \psi(c(x, \gamma)).$$

Then $\tilde\psi$ is conditionally negative definite on Γ.

Proof. For $\gamma, \gamma' \in \Gamma$ and $x \in X$, we have

$$c(x, \gamma^{-1}\gamma') = c(\gamma x, \gamma)^{-1}c(\gamma x, \gamma')$$

hence

$$\tilde\psi(\gamma^{-1}\gamma') = \sum_{x \in X} \psi(c(x, \gamma)^{-1}c(x, \gamma'));$$

but it is clear that, for all $x \in X$, the kernel

$$(\gamma, \gamma') \mapsto \psi(c(x, \gamma)^{-1}c(x, \gamma'))$$

is conditionally negative definite on Γ. $\qquad\square$

The first part of the following result was obtained in [Jol00], Proposition 2.5, with a different proof.

Proposition 6.2.3. *Let G, H be discrete groups containing a common subgroup A, and let $\Gamma = G \star_A H$ be the corresponding amalgamated product.*

(1) *If G and H have the Haagerup property and if A is finite, then Γ also has the Haagerup property.*

(2) *If A is central in G and in H and if the pairs (G, A) and (H, A) have the generalized Haagerup property, then so does the pair (Γ, A). In particular, Γ has the Haagerup property.*

Proof. First we prove part (1). Let R and S be sets of representatives for the left cosets of A in G and H respectively, such that $1 \in R$ and $1 \in S$. We denote elements of R, S and A by α_i, β_j and a respectively. According to [Ser77], every element $\gamma \in \Gamma$ has a normal form:

$$\gamma = \alpha_1 \beta_1 \alpha_2 \beta_2 \ldots \alpha_l \beta_l a. \tag{6.7}$$

We claim that, if ψ is a A-bi-invariant, conditionally negative definite function on H, then the function $\tilde{\psi}$ on Γ given by

$$\tilde{\psi}(\gamma) = \sum_{j=1}^{l} \psi(\beta_j),$$

where $\gamma \in \Gamma$ of the form (6.7), is conditionally negative definite.

To prove this claim, we remark that Γ/H may be identified, thanks to the normal form (6.7), with the set of words of the form $\alpha_1 \beta_1 \alpha_2 \beta_2 \ldots \alpha_k$ (with $\alpha_i \in R$ and $\beta_j \in S$); this provides a section $\sigma \colon \Gamma/H \to \Gamma$ for the canonical projection $\Gamma \to \Gamma/H$. Then we define $c \colon \Gamma/H \times \Gamma \to H$ by

$$c(x, \gamma) = \sigma(x)^{-1} \gamma \sigma(\gamma^{-1} x);$$

this is a cocycle in the sense of Lemma 6.2.2. Fix $\gamma \in \Gamma$ and $x \in \Gamma/H$.

If γ does not begin with the word $\sigma(x)$, we write $\sigma(x) = x_0 x_1$ and $\gamma = x_0 y_1 a$, where x_0 is the subword common to $\sigma(x)$ and γ, and x_1 ends with α_k. Then

$$\gamma^{-1} \sigma(x) = a^{-1} y_1^{-1} x_1 = y_1' x_1' a',$$

where x_1' ends with α_k'. Hence $\sigma(\gamma^{-1} x) = y_1' x_1'$ and $c(x, \gamma) = a'^{-1} \in A$. This already shows that, for fixed γ, the set $\{x \in \Gamma/H : c(x, \gamma) \notin A\}$ is finite.

We assume now that γ begins with the word $\sigma(x)$ and we compute $c(x, \gamma)$. The normal form of γ is

$$\gamma = \sigma(x) \beta_k \alpha_{k+1} \beta_{k+1} \ldots \beta_l a,$$

so

$$\gamma^{-1}\sigma(x) = a^{-1}\beta_l^{-1} \ldots \alpha_{k+1}^{-1}\beta_k^{-1} = \beta_l'\alpha_l' \ldots \beta_{k+1}'\alpha_{k+1}'a'\beta_k^{-1}$$

and

$$\sigma(\gamma^{-1}x) = \beta_l'\alpha_l' \ldots \beta_{k+1}'\alpha_{k+1}',$$

so that $c(x,\gamma) = \beta_k a'^{-1}$ and $\psi(c(x,\gamma)) = \psi(\beta_k)$. Consequently, if $\tilde{\psi}$ is the conditionally negative definite function defined on Γ as in Lemma 6.2.2, we indeed have $\tilde{\psi}(\gamma) = \sum_{j=1}^l \psi(\beta_j)$, which establishes the claim.

Now we prove the first part of Proposition 6.2.3 itself. Let ϕ and ψ be proper, conditionally negative definite functions on G and H respectively. By perturbing ϕ and ψ by bounded functions if necessary, we may assume that ϕ and ψ satisfy conditions (a) and (b) of Lemma 6.2.1, with respect to the finite subgroup A. The claim above then provides conditionally negative definite functions $\tilde{\phi}$ and $\tilde{\psi}$ on Γ. Write F for $\tilde{\phi} + \tilde{\psi}$; this is a conditionally negative definite function on Γ, and it remains to check that F is proper. For that, we fix a real number $T > 0$ and show that, if $F(\gamma) \leq T$, there are only finitely many possibilities for the normal form (6.7) of γ. As $\tilde{\psi}(\gamma) = \sum_{j=1}^l \psi(\beta_j) \geq l-1$, we first see that the length of the normal form is bounded. Next, since $\phi(\alpha_i) \leq T$ and $\psi(\beta_j) \leq T$, we deduce that there are finitely many choices for α_i and β_j. Since A is finite by assumption, the proof of the first part of Proposition 6.2.3 is concluded.

S.R. Gal [Gal] has generalized part (1) of Proposition 6.2.3 to certain amalgamated products $G *_A H$, with infinite A.

The proof of part (2) relies on the following result, which is proved in the appendix to this chapter; we still denote by R and S sets of representatives of the left cosets of A in G and H respectively, with $1 \in R$ and $1 \in S$. Note that A is central in Γ since the image of A in Γ commutes with both G and H.

Proposition 6.2.4. *Assume that A is central in both G and H and that $\chi \in \hat{A}$. If ϕ_χ and ψ_χ are normalized positive definite functions on G and H respectively, satisfying the covariance condition (2) of Definition 4.2.1, define ω_χ on Γ by*

$$\omega_\chi(\alpha_1\beta_1 \ldots \alpha_l\beta_l a) = \phi_\chi(\alpha_1)\psi_\chi(\beta_1)\ldots\phi_\chi(\alpha_l)\psi_\chi(\beta_l)\chi(a),$$

for every reduced word $\alpha_1\beta_1 \ldots \alpha_l\beta_l a \in \Gamma$, with $\alpha_j \in R$, $\beta_j \in S$ and $a \in A$ as above. Then ω_χ does not depend on the choice of sets of representatives R and S. Further, ω_χ is normalized and positive definite and satisfies the covariance condition

$$\omega_\chi(xa) = \omega_\chi(x)\chi(a) \qquad \forall x \in \Gamma \quad \forall a \in A.$$

Given this proposition, the proof of part (2) of Proposition 6.2.3 concludes as follows. Fix a finite subset K of Γ and $\varepsilon > 0$. There exist finite sets $F_1 \subset R$,

$F_2 \subset S$, $F \subset A$ and a positive integer N such that

$$K \subset \{\alpha_1 \beta_1 \ldots \alpha_l \beta_l a : l \leq N, \ \alpha_j \in F_1, \ \beta_j \in F_2, \ a \in F\}.$$

Choose $\delta \in \,]0, \varepsilon/(4N+1)[$. As (G, A) and (H, A) have the generalized Haagerup property, by part (2) of Lemma 6.2.1 there exist neighbourhoods V and V' of 1 in $\hat{A}$, families $\{\phi_\chi : \chi \in V\}$ and $\{\psi_\chi : \chi \in V'\}$ and a constant $c \in \,]0, 1[$ satisfying

(1) $|\phi_\chi(x) - 1| < \delta$ for all $x \in F_1$ and $\chi \in V$,
(2) $\phi_\chi(xa) = \phi_\chi(x)\chi(a)$ for all $x \in G$, $a \in A$ and $\chi \in V$,
(3) $\chi \mapsto \phi_\chi(x)$ is continuous, for all $x \in G$,
(4) $|\phi_\chi|$ belongs to $C_0(G/A)$, for all $\chi \in V$,
(5) $|\phi_\chi(g)| < c$ for all $g \in G \setminus A$,

and the analogous conditions for $\{\psi_\chi : \chi \in V'\}$ on H with respect to F_2 and δ. Let W be the set of all χ in $V \cap V'$ such that $|\chi(a) - 1| < \delta$ for all $a \in F$, which is a neighbourhood of 1 in $\hat{A}$. Then for $\chi \in W$, let ω_χ be the normalized positive definite function on Γ given by Proposition 6.2.4: it obviously fulfills conditions (2) and (3).

We now check that ω_χ satisfies condition (1) for every $\chi \in W$, with respect to K and ε. If $x = \alpha_1 \beta_1 \ldots \alpha_l \beta_l a \in K$ is a reduced word, then

$$|\omega_\chi(x) - 1| = |\phi_\chi(\alpha_1)\psi_\chi(\beta_1) \ldots \phi_\chi(\alpha_l)\psi_\chi(\beta_l)\chi(a) - 1|$$

$$\leq \sum_{j=1}^{l} (|\phi_\chi(\alpha_j) - 1| + |\psi_\chi(\beta_j) - 1|) + |\chi(a) - 1|$$

$$< (4l + 1)\delta \leq (4N + 1)\delta \leq \varepsilon.$$

Finally, we show that ω_χ satisfies condition (4) for $\chi \in W$: we have to prove that, for every $\eta > 0$, the subset

$$L = \{\dot{x} \in \Gamma/A : |\omega_\chi(\dot{x})| \geq \eta\}$$

is finite. To do this, take $\dot{x} = \alpha_1 \beta_1 \ldots \alpha_l \beta_l A \in L$, $\dot{x} \neq 1$. Then

$$|\omega_\chi(\dot{x})| = |\phi_\chi(\alpha_1)| \, |\psi_\chi(\beta_1)| \ldots |\phi_\chi(\alpha_l)| \, |\psi_\chi(\beta_l)| \geq \eta.$$

As all α_j and β_j are different from 1, except perhaps α_1 or β_l and as $|\phi_\chi(\alpha_j)| < c$ and $|\psi_\chi(\beta_j)| < c$ for these j, it follows that $\eta < c^{2l-1}$, hence the length of the normal form of all elements of L is smaller than some constant $M = M(t, \eta)$. Furthermore, the sets $\{r \in R : |\phi_\chi(r)| \geq \eta\}$ and $\{s \in S : |\psi_\chi(s)| \geq \eta\}$ are finite, proving that L is finite.

The remainder of Proposition 6.2.3 now follows from Lemma 4.2.11. $\square$

Corollary 6.2.5. *Let G and H be discrete amenable groups, and let A be a common central subgroup. Then the amalgamated product $\Gamma = G \star_A H$ has the Haagerup property.*

Proof. Combine Proposition 4.2.12 with part (2) of Proposition 6.2.3. $\qquad\square$

Example 6.2.6 (Torus knot groups). Fix integers $p, q \geq 2$, and let $\Gamma_{p,q}$ be the torus knot group $\langle x, y \mid x^p = y^q \rangle$. Since $\Gamma \simeq \mathbb{Z} \star_{\mathbb{Z}} \mathbb{Z}$, where the amalgamated $\mathbb{Z}$ embeds as $p\mathbb{Z}$ in the first factor and as $q\mathbb{Z}$ in the second factor, Corollary 6.2.5 applies to show that $\Gamma_{p,q}$ has the Haagerup property (see [BCS00] and Example 7.3.4 below for different proofs of the same fact).

For Higman–Neumann–Neumann extensions, we have a result analogous to the first part of Proposition 6.2.3 for amalgamated products.

Proposition 6.2.7. *Let H be a discrete group, let A be a finite subgroup and let $\theta \colon A \to H$ be a monomorphism. If H has the Haagerup property, then the Higman–Neumann–Neumann extension $\Gamma = \mathrm{HNN}(H, A, \theta)$ has the Haagerup property.*

Proof. By [Ser77, I.1.4, Prop. 5], Γ is a semidirect product $\Gamma = G \rtimes \mathbb{Z}$, where G is obtained by amalgamation of

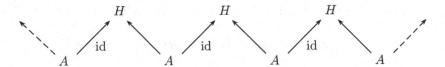

and $\mathbb{Z}$ acts by the shift. In other words, setting $H_n = H$ for every $n \in \mathbb{Z}$, the group G is the fundamental group of the graph of groups:

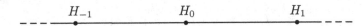

(see [Bau93, p. 134]); thus G is obtained as the inductive limit of the sequence $H_0 \star_A H_1$, $H_{-1} \star_A (H_0 \star_A H_1)$, $(H_{-1} \star_A (H_0 \star_A H_1)) \star_A H_2$, $\cdots$. It follows from the first part of Proposition 6.2.3 that each of these groups has the Haagerup property. Therefore, so does have G, by Proposition 6.1.1. Finally, since $\mathbb{Z}$ is amenable, $\Gamma = G \rtimes \mathbb{Z}$ has the Haagerup property, by Example 6.1.6. $\qquad\square$

Theorem 6.2.8. *Let Γ be a countable group acting on a tree without inversion, with finite edge stabilizers. If the vertex stabilizers in Γ have the Haagerup property, then so does Γ.*

Proof. By Bass–Serre theory ([Ser77, I.5.4, Thm 13]), Γ is the fundamental group of a graph of groups $(\mathcal{G}, Y)$ (where Y is a graph and $\mathcal{G}$ a system of groups, such that the edge groups are finite, and the vertex groups have the Haagerup property). We consider several cases.

First, if Y is a segment, then Γ is an amalgamated product, and part (1) of Proposition 6.2.3 applies.

Next, if Y is a loop, then Γ is a Higman–Neumann–Neumann extension: $\Gamma = \mathrm{HNN}(H, A, \theta)$, where A is a finite subgroup of H, and $\theta: A \to H$ is a monomorphism; Proposition 6.2.7 applies.

Third, if Y is finite, we argue by induction on the number n of edges of Y. If $n = 0$, there is nothing to prove. If $n > 0$, we choose some edge e and contract it; this does not change the fundamental group of the graph of groups ([Ser77, I.5.2, Lem. 6]). If e is a segment, with vertex groups G and H and edge group A, we replace e by one vertex, with $G \star_A H$ as associated group. If e is a loop, with vertex group H and edge group A, we replace the vertex group by $\mathrm{HNN}(H, A, \theta)$. In either case, we obtain a graph of groups with $n - 1$ edges, and the two previous cases show that induction applies to this graph of groups.

Finally, in general, a graph of groups $(\mathcal{G}, Y)$ is the inductive limit of its finite subgraphs, so that Proposition 6.1.1 allows one to conclude. $\qquad\square$

6.3 Group presentations

Let Γ be the finitely presented group $\langle X \mid r_1, r_2, \ldots, r_m \rangle$; we assume that the r_i are cyclically reduced words in the free group $F(X)$ with basis X. We shall denote by $\hat{r}_i$ the cyclic word obtained by closing r_i; we set $\hat{R} = \{\hat{r}_1, \ldots, \hat{r}_m\}$.

Theorem 6.3.1. *Let Γ be the finitely presented group $\langle X \mid r_1, r_2, \ldots, r_m \rangle$, and assume that*

(1) *$|r_i|_X \geq 6$ for $i = 1, \ldots, m$;*

(2) *every subword of length 2 in $F(X)$ occurs at most once in $\hat{R}$;*

(3) *if a subword of length 2 in $F(X)$ occurs in $\hat{R}$, then its inverse does not occur.*

Then Γ has the Haagerup property, and its classifying space is a finite complex of dimension 2.

Proof. Let Δ be the 2-complex associated with the presentation ([Mas67, Chap. 7, Cor. 2.2]): the 1-skeleton is a bouquet of $\mathrm{card}(X)$ circles, to which we attach m cells of dimension 2, according to the relations $r_1, \ldots, r_m$. The fundamental group of Δ is Γ. We denote by $\tilde{\Delta}$ the universal cover of Δ.

For an arbitrary finite presentation, we describe $\tilde{\Delta}$ more precisely. It is a polygonal 2-complex on which Γ acts freely. The numbers of sides of faces of $\tilde{\Delta}$ correspond to the lengths of the relations in the presentation. Moreover, the links of vertices of $\tilde{\Delta}$ are isomorphic to the link L of the unique vertex of Δ. Let us describe L as a graph. The set of vertices of L is $X \cup X^{-1}$. For $a, b \in X \cup X^{-1}$, we draw an edge in L between a^{-1} and b for each occurrence either of ab or of $b^{-1}a^{-1}$ in $\hat{R}$ (see the figure below). In other words, we draw an edge between a and b for each occurrence of $a^{-1}b$ or $b^{-1}a$ in $\hat{R}$ (notice that L has no loops). Consequently, the graph L is simple (that is, there is at most one edge between any two vertices) if and only if the girth of L is at least 3, and this holds if and only if the presentation fulfills conditions (2) and (3) above.

From [BŚ97] (see also Example 1.2.7) we recall that a polygonal 2-complex is a (k, l)-**complex** if every face has at least $k \geq 3$ sides, and if the girth of the link of every vertex is at least $l \geq 3$. If the presentation satisfies conditions (1), (2), (3) above, then $\tilde{\Delta}$ is a $(6, 3)$-complex. Since a simply connected $(6, 3)$-complex is a Hadamard space, hence is contractible (see [Bal95]), we see that Δ is a classifying space for Γ. Moreover, it follows from the results of W. Ballmann and J. Świątkowski [BŚ97] mentioned in Example 1.2.7 that a group acting properly on a simply connected $(6, 3)$-complex has the Haagerup property; this applies to Γ. $\square$

In the above proof, the total number of edges of the link L is $\sum_{i=1}^{m} |r_i|_X$. Set $k = \mathrm{card}(X)$. Conditions (2) and (3) of Theorem 6.3.1 imply that L is a simple graph, hence

$$6m \leq \sum_{i=1}^{m} |r_i|_X \leq k(2k - 1),$$

where $k(2k - 1)$ is the number of edges of the complete graph K_{2k} on $2k$ vertices. We notice that the upper bound is optimal. Indeed, if $k \geq 2$, the group generated by $x_1, \ldots, x_k$ with the relations

$$x_{k-1}x_k^2 x_{k-1}x_k^{-1}x_{k-1} \qquad \text{and} \qquad (\textstyle\prod_{j=i+1}^{k} x_i x_j x_i x_j^{-1})x_i \quad i = 1, \ldots, k - 2$$

has a link L isomorphic to K_{2k}. Since the relations have length at least 6, Theorem 6.3.1 applies, and Γ has the Haagerup property.

If $g \geq 1$, consider the surface group Γ_g, that is, the fundamental group of a closed Riemann surface of genus g, in its standard presentation:

$$\Gamma_g = \langle a_1, b_1, \ldots, a_g, b_g \mid \prod_{i=1}^{g} [a_i, b_i] \rangle.$$

If $g \geq 2$, then Theorem 6.3.1 applies to give a combinatorial proof of the Haagerup property for Γ_g.

To treat $\Gamma_1 \simeq \mathbb{Z}^2$, the following observation is necessary.

Let Γ be the finitely presented group $\langle X \mid r_1, \ldots, r_m \rangle$, and assume that

(1) $|r_i|_X \geq 4$ for $i = 1, \ldots, m$;

(2) every subword of length 2 in $F(X)$ occurs at most once in $\hat{R}$, and if such a subword occurs, then its inverse does not;

(3) for every 3-element subset F in $X \cup X^{-1}$, there exist $x, y \in F$ such that $x \neq y$ and $x^{-1}y$ and $y^{-1}x$ do not occur in $\hat{R}$.

Then the conclusions of Theorem 6.3.1 hold. Indeed, condition (3) ensures that the link L contains no triangles; combined with condition (2), this means that the girth of L is at least 4. In other words, $\tilde{\Delta}$ is a $(4,4)$-complex; but the results of Ballmann and Świątkowski [BŚ97] (see also Example 1.2.7) are valid for $(4,4)$-complexes.

In this situation, it is easy to see that L is bipartite on X and X^{-1} if and only if, in each relation r_i, all the exponents have the same sign. In particular, if all r_i are *positive* words, then condition (3) is automatic. A thorough study of one-relator groups with a positive relation appears in [Bau71].

Finally, it follows from Theorem E in [BB95] that the following version of the Tits' alternative holds for the groups discussed in Theorem 6.3.1 or satisfying the conditions (1)–(3) on the previous page: either they contain a free nonabelian subgroup, or they are isomorphic to a Bieberbach group of rank 2. In any case, these groups are infinite. However, it may happen that their abelianizations are finite. An example is provided by the presentation

$$\Gamma = \langle a, b, c, d \mid abacad = bcbdba^{-1} = cdca^{-1}cb^{-1} = da^{-1}db^{-1}dc^{-1} = 1 \rangle,$$

to which Theorem 6.3.1 applies.

6.4 Appendix: Completely positive maps on amalgamated products,
by Paul Jolissaint

Our aim is to give a proof of Proposition 6.2.4. It is inspired by the main theorem of a paper of F. Boca [Boc91].

Let $(A_i)_{i \in I}$ be a family of unital C*-algebras such that

(1) all A_i contain a common unital central C*-subalgebra B;

(2) for every $i \in I$, there exists a conditional expectation $E_i \colon A_i \to B$.

Set $A_i^0 = \ker(E_i)$, so that $A_i = B \oplus A_i^0$ as a B-bimodule. For a positive integer n, define D_1 to be I and, for $n \geq 2$, D_n by the rule

$$D_n = \{\mathbf{i} \in I^n : i_j \neq i_{j+1}\}$$

(we use multi-index notation, so that $\mathbf{i}$ stands for $(i_1, \dots, i_n)$). Then the **(algebraic) amalgamated product** of the A_i over B is denoted by $\star_B A_i$. As a B-bimodule, it is equal to

$$B \oplus \bigoplus_{\substack{n \geq 1 \\ \mathbf{i} \in D_n}} A_{i_1}^0 \otimes_B \dots \otimes_B A_{i_n}^0.$$

Equipped with the involution defined by the formula $(c_1 \dots c_n)^* = c_n^* \dots c_1^*$, for every reduced word $c_1 \dots c_n \in A_{i_1}^0 \otimes_B \dots \otimes_B A_{i_n}^0$, $A = \star_B A_i$ is a unital *-algebra.

Theorem 6.4.1. *Let C be a unital C*-algebra. Assume that there exist a unital *-homomorphism π from B to the center of C and unital completely positive maps $\Phi_i \colon A_i \to C$ for all $i \in I$, satisfying the covariance condition*

$$\Phi_i(ab) = \Phi_i(a)\pi(b)$$

for all $a \in A_i$ and $b \in B$. Define $\Phi \colon \star_B A_i \to C$ by $\Phi|_B = \pi$ and, for $\mathbf{i} \in D_n$ and $c_1 \dots c_n \in A_{i_1}^0 \otimes_B \dots \otimes_B A_{i_n}^0$,

$$\Phi(c_1 \dots c_n) = \Phi_{i_1}(c_1) \dots \Phi_{i_n}(c_n).$$

Then Φ is unital and completely positive.

Proof. We first fix notation and recall a few facts needed for the proof.

Write W for the set of **words** in A:

$$W = B \cup \{c_1 \dots c_n : n \geq 1, \ c_k \in A_{i_k}^0, \ \mathbf{i} \in D_n\},$$

and W_0 for $(W \setminus B) \cup \{1\}$. For $x \in W$, its **height** $h(x)$ is defined to be 0 if $x \in B$ and to be n if $x = c_1 \dots c_n \in A_{i_1}^0 \otimes_B \dots \otimes_B A_{i_n}^0$ with $(i_1, \dots, i_n) \in D_n$; the multi-index $\mathbf{i}$ is called the **type** of x. If $w = c_1 \dots c_n \in W$, we write $\tilde{w}$ for $\{1, c_1, c_1 c_2, \dots, w\}$, and we say that a subset X of W_0 is **complete** if $1 \in X$ and if $\tilde{w} \subset X$ for every $w \in X$.

For $\mathbf{i} = (i_1, \dots, i_n) \in D_n$, let $A^0_{\mathbf{i}}$ denote

$$A^0_{i_1}$$
$$\oplus A^0_{i_1} \otimes_B A^0_{i_2} \otimes_B A^0_{i_1}$$
$$\oplus A^0_{i_1} \otimes_B A^0_{i_2} \otimes_B A^0_{i_3} \otimes_B A^0_{i_2} \otimes_B A^0_{i_1}$$
$$\oplus \dots$$
$$\oplus A^0_{i_1} \otimes_B \dots \otimes_B A^0_{i_n} \otimes_B \dots \otimes_B A^0_{i_1},$$

and finally, for $i \in I$, let $A_l(i)$ denote

$$B \oplus \bigoplus_{\substack{n \geq 1 \\ i \in D_n \\ i_1 \neq i}} A^0_{i_1} \otimes_B \dots \otimes_B A^0_{i_n}.$$

We take a Hilbert space $\mathcal{H}$ on which the C*-algebra C acts faithfully. Let X be a finite set, let D be a C*-algebra and let $k \colon X \times X \to D$ be a positive definite kernel: this means that the matrix $(k(x, y))_{x,y \in X} \in M_n(D)$ is positive, where $n = \operatorname{card}(X)$. Then for every $w \in X$, there exists $v_w \colon X \to D$ such that

$$k(x, y) = \sum_{w \in X} v_w(y)^* v_w(x),$$

for all $x, y \in X$. Observe that, if $v(x) = (v_w(x))_{w \in X}$ is considered as a column matrix, we may also write $k(x, y) = v(y)^* v(x)$.

We are now ready to prove the theorem. As in the proof of [Boc91, Prop. 3.2], we must show that for every complete finite subset X of W_0 and every map $f \colon X \to \mathcal{H}$, the double sum

$$S_X(f) = \sum_{x,y \in X} \langle \Phi(y^* x) f(x), f(y) \rangle \geq 0.$$

We proceed by induction on $\operatorname{card}(X)$. If $\operatorname{card}(X) = 1$ or 2, then $S_X(f) \geq 0$ because $X = \{1\}$ in the first case and $X = \{1, a\}$ with $a \in A^0_i$ for some $i \in I$ in the second, and $\Phi|_{A_i} = \Phi_i$.

Now, let $X \subset W_0$ be finite and complete such that the desired result is true for every complete $Y \subset W_0$ with $\operatorname{card}(Y) < \operatorname{card}(X)$. Let also $f \colon X \to \mathcal{H}$ be as above. Set $m = \max_{x \in X} h(x)$ and choose some $w_0 = a_1 \dots a_m \in X$ with $a_j \in A^0_{i_j}$, $\mathbf{i} \in D_m$. Then define X_2 to be the set of all $x \in X$ of the same type as w_0, that is, the set of all elements of the form $c_1 \dots c_m$ where $c_j \in A^0_{i_j}$, and define X_1 to be $X \setminus X_2$, which is complete. Then $X = X_1 \sqcup X_2$, and by the induction hypothesis, the kernel $(x, y) \mapsto \Phi(y^* x)$ on $X_1 \times X_1$ is positive definite, hence we may write

$$\Phi(y^* x) = v(y)^* v(x) \qquad \forall x, y \in X.$$

Next we decompose $S_X(f)$:

$$S_X(f) = \sum_{x,y \in X_1} \langle \Phi(y^*x)f(x), f(y) \rangle$$

$$+ \sum_{x_0a, x_0'a' \in X_2} \langle \Phi(a'^*x_0'^*x_0a)f(x_0a), f(x_0'a') \rangle$$

$$+ 2\,\mathrm{Re} \sum_{x \in X_1, x_0a \in X_2} \langle \Phi(a^*x_0^*x)f(x), f(x_0a) \rangle,$$

since every $x_2 \in X_2$ is of the form $x_2 = x_0a$ where $x_0 = c_1 \ldots c_{m-1}$, with each $c_k \in A_{i_k}^0$, and $a \in A_{i_m}^0$. Observe that x_0^*x belongs to $A_l(i_m)$ for x_0 as above and $x \in X_1$, hence

$$\Phi(a^*x_0^*x) = \Phi_{i_m}(a^*)\Phi(x_0^*x) = \Phi(a^*)\Phi(x_0^*x).$$

Now let X_0 be the set of all $c_1 \ldots c_{m-1}$, where each $c_k \in A_{i_k}^0$, such that there exists some $a \in A_{i_m}^0$ such that $c_1 \ldots c_{m-1}a \in X_2$. Then $X_0 \subseteq X_1$.

Again by the proof of [Boc91, Prop. 3.2], for every pair $(x_0, x_0') \in X_0 \times X_0$,

$$x_0'^*x_0 = b(x_0, x_0') + d(x_0, x_0')$$

where $b \colon X_0 \times X_0 \to B$ is a positive definite kernel and $d(x_0, x_0') \in A_{(i_{m-1}, \ldots, i_1)}^0$. Thus we may write

$$b(x_0, x_0') = \sum_{w \in X_0} \beta_w(x_0')^*\beta_w(x_0)$$

where $\beta_w(x_0) \in B$ for all $w, x_0 \in X_0$. Moreover, by Stinespring's theorem applied to Φ_{i_m}, there exist a unital *-homomorphism $\rho \colon A_{i_m} \to \mathcal{L}(\mathcal{K})$ and an isometry $V \colon \mathcal{H} \to \mathcal{K}$ such that

$$\Phi_{i_m}(a) = V^*\rho(a)V \qquad \forall a \in A_{i_m}.$$

For $x_0a, x_0'a' \in X_2$, it follows that $\Phi(a'^*x_0'^*x_0a)$ is equal to

$$\Phi(a'^*b(x_0, x_0')a) + \Phi(a'^*d(x_0, x_0')a)$$

$$= \Phi(a'^*b(x_0, x_0')a) + \Phi(a'^*)\Phi(d(x_0, x_0'))\Phi(a)$$

$$= \Phi(a'^*b(x_0, x_0')a) + \Phi(a'^*)\Phi(x_0'^*x_0)\Phi(a) - \Phi(a'^*)\pi(b(x_0, x_0'))\Phi(a)$$

$$= \Phi_{i_m}(a'^*b(x_0, x_0')a) + \Phi(a'^*)v_{x_0'}^*v_{x_0}\Phi(a) - \Phi_{i_m}(a'^*)\pi(b(x_0, x_0'))\Phi_{i_m}(a)$$

$$= \sum_{w \in X_0} \pi(\beta_w(x_0')^*)[\Phi_{i_m}(a'^*a) - \Phi_{i_m}(a')^*\Phi_{i_m}(a)]\pi(\beta_w(x_0))$$

$$+ \Phi(a')^*v_{x_0'}^*v_{x_0}\Phi(a),$$

by the construction, the facts that $d(x_0, x_0') \in A^0_{(i_{m-1}, \ldots, i_1)}$ and $a, a' \in A^0_{i_m}$, the inductive hypothesis, and the fact that $x_0, x_0' \in X_1$.

Setting $g(x_0, a, w) = \pi(\beta_w(x_0))f(x_0 a)$, we see that

$$\sum_{x_0 a, x_0' a' \in X_2} \left\langle \Phi(a'^* x_0'^* x_0 a)f(x_0 a), f(x_0' a') \right\rangle$$

$$= \sum_{w, x_0 a, x_0' a'} \left\langle (V^* \rho(a'^* a)V - V^* \rho(a'^*)VV^* \rho(a)V)g(x_0, a, w), g(x_0', a', w') \right\rangle$$

$$+ \sum_{x_0 a, x_0' a'} \left\langle v_{x_0} \Phi(a)f(x_0 a), v_{x_0'} \Phi(a')f(x_0' a') \right\rangle$$

$$= \sum_{w, x_0 a, x_0' a'} \left\langle (1 - VV^*)\rho(a)Vg(x_0, a, w), \rho(a')Vg(x_0', a', w) \right\rangle$$

$$+ \left\| \sum_{x_0 a} v_{x_0} \Phi(a)f(x_0 a) \right\|^2$$

$$\geq \left\| \sum_{x_0 a} v_{x_0} \Phi(a)f(x_0 a) \right\|^2.$$

We deduce from this that

$$S_X(f) \geq \left\| \sum_{x \in X_1} v_x f(x) \right\|^2 + \left\| \sum_{x_0 a \in X_2} v_{x_0} \Phi(a)f(x_0 a) \right\|^2$$

$$+ 2 \operatorname{Re} \sum_{x \in X_1, x_0 a \in X_2} \left\langle \Phi(a)^* \Phi(x_0^* x)f(x), f(x_0 a) \right\rangle$$

$$= \left\| \sum_{x \in X_1} v_x f(x) \right\|^2 + \left\| \sum_{x_0 a \in X_2} v_{x_0} \Phi(a)f(x_0 a) \right\|^2$$

$$+ 2 \operatorname{Re} \sum_{x \in X_1, x_0 a \in X_2} \left\langle v_x f(x), v_{x_0} \Phi(a)f(x_0 a) \right\rangle$$

$$= \left\| \sum_{x \in X_1} v_x f(x) + \sum_{x_0 a \in X_2} v_{x_0} \Phi(a)f(x_0 a) \right\|^2 \geq 0.$$

This ends the proof of Theorem 6.4.1. $\qquad\square$

Before proving Proposition 6.2.4, recall that if φ is a normalized positive definite function on a discrete group Γ, then φ defines a unital completely positive map m_φ (called a multiplier) on the reduced C*-algebra $C_r^*(\Gamma)$, characterized by

$$m_\varphi(\lambda(g)) = \varphi(g)\lambda(g) \qquad \forall g \in \Gamma.$$

Conversely, if Φ is a unital completely positive map on $C_r^*(\Gamma)$ such that, for every $g \in \Gamma$, $\Phi(\lambda(g)) = \varphi(g)\lambda(g)$ for some scalar $\varphi(g)$, then φ is a normalized positive definite function which is recaptured by $\varphi(g) = \tau(\Phi(\lambda(g))\lambda(g)^{-1})$, where τ is the canonical trace on $C_r^*(\Gamma)$.

Proof of Proposition 6.2.4. As the character χ is fixed, we set $\phi = \phi_\chi$ and $\psi = \psi_\chi$. We check first that ω_χ does not depend on the choices of the sets of representatives of A-cosets $R \subset G$ and $S \subset H$. Indeed, if R' and S' are other such sets, and $x = \alpha_1\beta_1 \ldots \alpha_l\beta_l a$ is a reduced word with $\alpha_j \in R$ and $\beta_j \in S$, as the type of x is independent of the chosen systems, we may write $x = \alpha_1'\beta_1' \ldots \alpha_l'\beta_l'a'$ with $\alpha_j' \in R'$ and $\beta_j' \in S'$. Thus for every j, there exist $a_j, b_j \in A$ such that $\alpha_j' = \alpha_j a_j$ and $\beta_j' = \beta_j b_j$ for every j. The uniqueness of the reduced form with respect to R and S shows that $a_1 b_1 \ldots a_l b_l a' = a$, and this in turn implies that

$$
\begin{aligned}
\phi(\alpha_1')&\psi(\beta_1') \ldots \phi(\alpha_l')\psi(\beta_l')\chi(a') \\
&= \phi(\alpha_1)\chi(a_1)\psi(\beta_1)\chi(b_1) \ldots \phi(\alpha_l)\chi(a_l)\psi(\beta_l)\chi(b_l)\chi(a') \\
&= \phi(\alpha_1) \ldots \psi(\beta_l)\chi(a_1 \ldots b_l a') \\
&= \phi(\alpha_1)\psi(\beta_1) \ldots \phi(\alpha_l)\psi(\beta_l)\chi(a),
\end{aligned}
$$

as required.

Consider the completely positive multipliers $\Phi = m_\phi$ and $\Psi = m_\psi$ on $C_r^*(G)$ and $C_r^*(H)$ respectively. Now $C_r^*(A)$ is a unital central C*-subalgebra of both $C_r^*(G)$ and $C_r^*(H)$, and the multiplier $E_A = m_{\chi_A}$ is a conditional expectation onto $C_r^*(A)$ in both algebras. Moreover, by assumption, the restrictions of Φ and Ψ to $C_r^*(A)$ coincide and define a *-homomorphism π, characterized by

$$\pi(\lambda(a)) = \chi(a)\,\lambda(a) \qquad \forall a \in A.$$

Denote by Ω the unital completely positive map given by Theorem 6.4.1 on the *-algebra $C = C_r^*(G) \star_{C_r^*(A)} C_r^*(H)$, which obviously contains a copy of $\Gamma = G \star_A H$. As the canonical traces on $C_r^*(G)$ and on $C_r^*(H)$ extend to a faithful normalized trace τ on C satisfying $\tau(g) = 0$ for every $g \neq 1$, we see that

$$\omega_\chi(g) = \tau(\Omega(g)g^{-1}) \qquad \forall g \in \Gamma,$$

which shows that ω_χ is positive definite. $\qquad\qquad\square$

Chapter 7

Open Questions and Partial Results
by Alain Valette

7.1 Obstructions to the Haagerup property

We have seen that, if a locally compact group G contains a closed noncompact subgroup H such that the pair (G, H) has relative property (T), then G may not have the Haagerup property. Hence we may ask whether relative property (T) (with respect to a noncompact subgroup) is the only obstruction to the Haagerup property. It was proved in Theorem 4.0.1 that the answer is positive for connected Lie groups[1].

7.2 Classes of groups

Here are some classes of discrete groups which are known *not* to contain any infinite subgroup with property (T), but for which the Haagerup property is unknown.

7.2.1 One-relator groups

Let $\Gamma = \langle X \mid r \rangle$ be a one-relator group. To avoid cyclic groups, we assume that the generating set X has at least 2 elements. Since Γ is K-amenable (see [BBV99]), a subgroup with property (T) is necessarily finite. Does Γ have the Haagerup property? The test case seems to be one-relator groups with torsion: these are known to admit a Dehn algorithm (see [LS77, Chap. IV,

---

[1]S. Popa suggested that, at the philosophical level, our question is analogous to von Neumann's question: "Is the presence of a free nonabelian subgroup the only obstruction to amenability?" ... and that both questions might share the same odd fate!

© Springer Basel 2001
P.-A. Cherix et al., *Groups with the Haagerup Property*,
Modern Birkhäuser Classics, DOI 10.1007/978-3-0348-0906-1_7

Thm 5.5]), hence to be hyperbolic in the sense of M. Gromov [Gro87]. The Haagerup property is known for a number of one-relator groups with geometric significance (for example, surface groups), and Theorem 6.3.1 provides more examples. Here are some partial results about one-relator groups.

First, if Γ is a one-relator group with nontrivial center, then Γ has the Haagerup property: this is due to C. Béguin and T. Ceccherini-Silberstein [BCS00]. This covers in particular the torus knot groups $\langle x, y \mid x^p = y^q \rangle$ (see Examples 6.2.6 and 7.3.4 for other proofs of the Haagerup property for these groups). Unfortunately, as discussed in [BCS00], the class of one-relator groups with nontrivial center is rather restricted; indeed, if $\Gamma = \langle X \mid r \rangle$ has nontrivial center, then card$(X) \leq 2$.

Now consider the "Baumslag–Solitar monsters" [BS62]; these are the groups

$$BS_{p,q} = \langle a, b \mid ab^p a^{-1} = b^q \rangle,$$

where $p, q \geq 1$, called monsters because $BS_{p,q}$ is nonhopfian when p and q are coprime and at least 2. It is a nice observation, due to S.R. Gal and T. Januszkiewicz [GJ00], that $BS_{p,q}$ has the Haagerup property. Indeed, on one hand $BS_{p,q}$ is the Higman–Neumann–Neumann extension HNN$(\mathbb{Z}, p\mathbb{Z}, \theta)$ with $\theta(p) = q$: this realizes $BS_{p,q}$ as a subgroup of Aut(T_{p+q}), where T_{p+q} is the homogeneous tree of degree $p + q$. On the other hand there is an noninjective homomorphism β from $BS_{p,q}$ to the affine group Aff$(\mathbb{R})$ of the real line, given by $\beta(a)(x) = qx/p$ and $\beta(b)(x) = x + 1$. As observed in [GJ00], the diagonal embedding $BS_{p,q} \to \text{Aut}(T_{p+q}) \times \text{Aff}(\mathbb{R})$ has a discrete image; since Aut(T_{p+q}) and Aff$(\mathbb{R})$ have the Haagerup property, by Examples 1.2.3 and 1.2.6, so has $BS_{p,q}$.

7.2.2   Three-manifold groups

Let M be a three-dimensional compact manifold, possibly with boundary. It follows from Thurston's geometrization conjecture (see for instance, [Thu82]) that the fundamental group $\pi_1(M)$ does not have property (T) unless it is finite. It would be interesting to have an unconditional proof of this statement. Of course one may ask a stronger question: does $\pi_1(M)$ have the Haagerup property? The question is already interesting for fundamental groups of special classes of 3-manifolds. Here we have in mind fundamental groups of Haken 3-manifolds (see [Hem76]); these groups are torsion-free and K-amenable (see [BBV99] for the latter), so that they do not contain any nontrivial subgroup with property (T); a subclass of Haken 3-manifolds groups is the class of knot groups (that is, fundamental groups of knot complements in S^3).

7.2.3 Braid groups

Denote by B_n the braid group on n strings; it may be presented as the group generated by $a_1, \ldots, a_{n-1}$, with the relations

$$\begin{cases} a_i a_{i+1} a_i = a_{i+1} a_i a_{i+1} & \text{if } 1 \leq i \leq n-2 \\ a_i a_j = a_j a_i & \text{if } |i-j| \geq 2 \end{cases}$$

Here again, this group is torsion-free and K-amenable (see [OO98] for the latter), so that it has no subgroups with property (T). Notice that the assumptions of Theorem 6.3.1 are not satisfied. So does B_n have the Haagerup property? The answer is positive for $n = 3$, since in this case the kernel of the homomorphism from B_3 to $\mathbb{Z}$ which maps a_1 and a_2 to 1 is a free group, and Proposition 6.1.5 applies.

Other classes of groups may be considered *ad libitum*.

7.3 Group constructions

What are the group constructions that preserve the class of groups with the Haagerup property? For instance, this class is stable under direct products (trivially), and under free products, or amalgamated products over finite subgroups (see Proposition 6.2.3).

7.3.1 Semi-direct products

The examples of $\mathbb{R}^2 \rtimes SL_2(\mathbb{R})$ and $\mathbb{Z}^2 \rtimes SL_2(\mathbb{Z})$ show that the class of groups with the Haagerup property is not closed under semidirect products. Let N and H be locally compact groups with the Haagerup property, with H acting continuously on N. It would be interesting to find conditions on the homomorphism from H to $\mathrm{Aut}N$ which ensure that the semidirect product $N \rtimes H$ has the Haagerup property.

7.3.2 Actions on trees

Let X be a tree. If a group Γ acts on X with finite edge stabilizers, and with vertex stabilizers having the Haagerup property, then Γ has the Haagerup property, by Theorem 6.2.8. This is not true anymore if edge stabilizers are only assumed to be amenable (or even abelian). Indeed, consider the two matrices

$$\begin{pmatrix} 1 & 2 \\ 0 & 1 \end{pmatrix} \quad \text{and} \quad \begin{pmatrix} 1 & 0 \\ 2 & 1 \end{pmatrix}; \tag{7.1}$$

it is well known that they generate a free group $\mathbb{F}_2$, of finite index in $\mathrm{SL}_2(\mathbb{Z})$. We denote by $\Gamma = \mathbb{Z}^2 \rtimes \mathbb{F}_2$ the corresponding semidirect product. We let $\mathbb{F}_2$ act on its Cayley tree, and view this action as an action of Γ via the surjective quotient map from Γ to $\mathbb{F}_2$. Both the edge and the vertex stabilizers in Γ are isomorphic to $\mathbb{Z}^2$. On the other hand, the pair $(\Gamma, \mathbb{Z}^2)$ has relative property (T); therefore Γ does not have the Haagerup property.

In view of this example, for a group Γ acting on a tree X with vertex stabilizers having the Haagerup property, we may ask whether Γ has the Haagerup property if the edge stabilizers are virtually cyclic. We may also ask whether one may relax the assumptions on the edge stabilizers in Theorem 6.2.8 by assuming that Γ acts effectively on X (that is, the homomorphism $\Gamma \to \mathrm{Aut}X$ is injective)? Another question is to give a proof of Theorem 6.2.8 using only the action of the group on the tree (that is, without using Bass–Serre theory).

7.3.3   Central extensions

Let $1 \to Z \to \tilde{G} \to G \to 1$ be a central extension of a locally compact group. Is it true that $\tilde{G}$ has the Haagerup property if and only if G has? It seems that neither of the two implications is obviously true or false. Some hope for positive answers is provided by Theorem 4.0.1, which implies that, if G and $\tilde{G}$ are connected Lie groups, then indeed $\tilde{G}$ has the Haagerup property if and only if G does. It is also relevant here to mention Lemma 4.2.8: if the pair $(\tilde{G}, Z)$ has the generalized Haagerup property, then G has the Haagerup property.

We now present a result allowing one to lift the Haagerup property from G to $\tilde{G}$.

Lemma 7.3.1. *Suppose that $1 \to Z \to \tilde{G} \xrightarrow{p} G \to 1$ is a central extension of locally compact groups, and that G has the Haagerup property. Assume that G contains a co-Følner subgroup H over which the central extension splits (that is, the central extension $1 \to Z \to p^{-1}(H) \xrightarrow{p} H \to 1$ splits). Then $\tilde{G}$ has the Haagerup property.*

Proof. Since the central sequence splits over H, the group $p^{-1}(H)$ is isomorphic to the direct product $Z \times H$, so it has the Haagerup property. On the other hand, since $p^{-1}(H)$ is co-Følner in $\tilde{G}$, Proposition 6.1.5 applies and $\tilde{G}$ has the Haagerup property.   □

Proposition 7.3.2. *Suppose that $1 \to Z \to \tilde{G} \xrightarrow{p} G \to 1$ is a central extension of locally compact groups, and that G has the Haagerup property. Assume that G contains a discrete co-Følner subgroup Γ such that $H^2(\Gamma, Z) = 0$ (where Z is a trivial Γ-module). Then $\tilde{G}$ has the Haagerup property.*

Proof. Using the well-known fact that $H^2(\Gamma, Z)$ parametrizes central extensions of Γ by Z, we see from the vanishing of $H^2(\Gamma, Z)$ that

$$1 \to Z \to \tilde{G} \to G \to 1$$

splits over Γ. The result then follows from Lemma 7.3.1. □

Recall that, if Γ is a free group, then $H^2(\Gamma, A) = 0$ for every Γ-module A. From this we immediately deduce the following result.

Corollary 7.3.3. *Let G be a locally compact group with the Haagerup property. Suppose that G contains a free, discrete, co-Følner subgroup. Then, for any central extension $1 \to Z \to \tilde{G} \to G \to 1$, the middle term $\tilde{G}$ has the Haagerup property.*

We give two examples where Corollary 7.3.3 applies.

Example 7.3.4 (Torus knot groups). Fix integers $p, q \geq 2$; the free product $(\mathbb{Z}/p\mathbb{Z}) \star (\mathbb{Z}/q\mathbb{Z})$ is virtually free. Let $\Gamma_{p,q} = \langle x, y \mid x^p = y^q \rangle$ be a torus knot group. Since $\Gamma_{p,q}$ appears as the middle term of a central extension

$$0 \to \mathbb{Z} \to \Gamma_{p,q} \to (\mathbb{Z}/p\mathbb{Z}) \star (\mathbb{Z}/q\mathbb{Z}) \to 1,$$

we see that $\Gamma_{p,q}$ has the Haagerup property (see [BCS00] for a different proof of the same fact).

Example 7.3.5. (The universal covering group of $SL_2(\mathbb{R})$*)* It is well known that $SL_2(\mathbb{R}) \simeq SU(1,1)$ admits the free group on two generators as a lattice (for example, the matrices in 7.1 generate a lattice in $SL_2(\mathbb{R})$). From Corollary 7.3.3, we see in particular that the universal covering group $\widetilde{SL_2}(\mathbb{R}) \simeq \widetilde{SU}(1,1)$ has the Haagerup property.

We recall from Proposition 4.2.14 and Lemma 4.2.11 that the universal covering group $\widetilde{SU}(n,1)$ of $SU(n,1)$ has the Haagerup property. So it is interesting to ask whether $SU(n,1)$ contain a discrete, co-Følner subgroup Γ for which $H^2(\Gamma, \mathbb{Z}) = 0$ $(n \geq 2)$. If so, this would provide, as above in the case where $n = 1$, a "soft" proof of the Haagerup property for the group $\widetilde{SU}(n,1)$. It was pointed out to us by B. Klingler and P. Pansu that, if such a Γ exists, it cannot be a lattice. Indeed, for Λ a lattice in $SU(n,1)$ with $n \geq 2$, the Kähler class on the corresponding Riemannian locally symmetric space defines a nonzero element in $H^2(\Lambda, \mathbb{R})$ (this is of course easy if Λ is cocompact).

7.4 Geometric characterizations

7.4.1 Chasles' relation

In Section 3.1, a proper isometric action of a rank 1 simple Lie group G on a metric space X (namely the Riemannian symmetric space of G) was exhibited, together with a G-equivariant proper cocycle on X taking values in a representation space of G. When this representation is unitary (that is, when $G = \mathrm{SO}(n,1)$ or $\mathrm{SU}(m,1)$), this implies that G has the Haagerup property. As noticed by Julg in [Jul98, 1.4], this situation can be turned into a geometric characterization of the Haagerup property.

Proposition 7.4.1. *Let G be a locally compact group. Then G has the Haagerup property if and only if there exists a metric space (X, d) on which G acts isometrically and metrically properly, a unitary representation π of G on a Hilbert space $\mathcal{H}_\pi$, and a continuous map $c \colon X \times X \to \mathcal{H}_\pi$ such that*

(1) *the **cocycle relation**, or **Chasles' relation**:*

$$c(x, z) = c(x, y) + c(y, z) \qquad \forall x, y, z \in X;$$

(2) *the **G-equivariance condition**:*

$$c(gx, gy) = \pi(g)c(x, y) \qquad \forall x, y \in X \quad \forall g \in G;$$

(3) *the **properness condition**:*

$$\|c(x, y)\| \to +\infty \qquad as \quad d(x, y) \to +\infty.$$

Proof. Assume that G has the Haagerup property, and let α be a metrically proper, affine, isometric action of G on a Hilbert space $\mathcal{H}$. It is enough to take for X the Hilbert space $\mathcal{H}$, for π the linear part of the action α, and for $c \colon \mathcal{H} \times \mathcal{H} \to \mathcal{H}$ the canonical cocycle $c \colon (\xi, \eta) \mapsto \xi - \eta$.

Now we show the converse. Fix $x_0 \in X$; then the action α of G on $\mathcal{H}_\pi$, defined by

$$\alpha(g) = \pi(g) + c(gx_0, x_0) \qquad \forall g \in G,$$

is affine and isometric, by conditions (1) and (2). This action is metrically proper, by the properness of c and of the action of G on X. $\qquad\qquad\square$

In [Jul98, 1.4], Julg revisits most of the examples listed in Chapter 1, and gives in each case a description of the space X and the cocycle c. He also uses Proposition 7.4.1 implicitly in Chapter 3, in his proof of the Haagerup property for $\mathrm{SO}(n,1)$ and $\mathrm{SU}(n,1)$.

As a corollary to Proposition 7.4.1, we give the proof of a result due to F. Haglund, F. Paulin and the author of this chapter (see also [Sha99]), already mentioned in Example 1.2.7.

Corollary 7.4.2. *Let $(X, \mathcal{W})$ be a space with walls. If the locally compact group G acts properly on $(X, \mathcal{W})$, then G has the Haagerup property.*

Proof. Recall that $w(x, y)$ denotes the number of walls separating x from y.

We first show that we may assume that $w(x, y) > 0$ unless $x = y$. Indeed, w satisfies the triangle inequality $w(x, z) \leq w(x, y) + w(y, z)$ for all x, y, z in X. We identify points x and y in X for which $w(x, y) = 0$ in the usual way, and denote by Y the quotient set. Walls of X become walls of Y, so Y becomes a space of walls on which G acts properly and isometrically, and further it is a discrete metric space with respect to w.

By replacing X by Y if necessary, we may and shall assume that w is a distance function on X. Now we construct a unitary representation π and a map $c \colon X \times X \to \mathcal{H}_\pi$ satisfying conditions (1)–(3) of Proposition 7.4.1. Define a half space to be one of the two classes in the partition of X defined by any wall. Let H be the set of half-spaces in X, and let π be the permutation representation of G on $\ell^2(H)$. For x in X, define χ_x to be the characteristic function of the set of half-spaces containing x. By the definition of a space with walls, for all x and y in X, the function $\chi_x - \chi_y$ has finite support in $\ell^2(H)$. We define $c(x, y)$ to be $\chi_x - \chi_y$, and observe that $\|c(x, y)\|^2 = 2w(x, y)$, so that c satisfies all three assumptions of the proposition. This concludes the proof. $\qquad\square$

7.4.2 Some cute and sexy spaces

Let $\mathcal{H}$ be a real, infinite-dimensional, separable Hilbert space, and denote by $O(\infty)$ its orthogonal group. Fix an orthonormal basis $(e_n)_{n \geq 1}$, and consider the quadratic form $-x_1^2 + \sum_{n=2}^\infty x_n^2$ on $\mathcal{H}$. Denote by $O(1, \infty)$ the group of bounded invertible operators on $\mathcal{H}$ preserving this quadratic form. The quotient

$$\mathbb{H}^\infty(\mathbb{R}) = O(1, \infty)/O(1) \times O(\infty)$$

is the **infinite-dimensional real hyperbolic space**. Similarly, using the complexification $\mathcal{H}_\mathbb{C}$, we define the **infinite-dimensional complex hyperbolic space** $\mathbb{H}^\infty(\mathbb{C})$; the corresponding group is $U(1, \infty)$. We treasure the following quotation from Gromov ([Gro93, 6.A.III]): "The spaces like these look as cute and sexy as their finite-dimensional brothers and sisters". Our goal here is to publicize the following result of Gromov ([Gro93, 7.A.III]).

Theorem 7.4.3. *Let G be a second countable, locally compact group. The following statements are equivalent:*

(1) *G has the Haagerup property;*

(2) *G admits a metrically proper isometric action on $\mathbb{H}^\infty(\mathbb{R})$;*

(3) *G admits a metrically proper isometric action on $\mathbb{H}^\infty(\mathbb{C})$.*

Proof. We show first that (1) implies (2). Suppose that G has the Haagerup property; then G admits a metrically proper, affine, isometric action on some *separable* real Hilbert space. To see this, let G act metrically properly, isometrically on some affine real Hilbert space, and consider the closed affine subspace generated by some G-orbit: since G is second countable, this is a separable, affine real Hilbert space.

To show that G admits a metrically proper isometric action on $\mathbb{H}^\infty(\mathbb{R})$, it is enough to see that the isometry group of a separable, affine real Hilbert space $\mathcal{K}$ embeds as a closed subgroup in $O(1, \infty)$. This is classical: in the decomposition $\mathcal{H} = \mathbb{R}e_1 \oplus \mathbb{R}e_2 \oplus \{e_1, e_2\}^\perp$, identify $\mathcal{K}$ with $\{e_1, e_2\}^\perp$; then, using the semidirect product decomposition $\mathrm{Isom}(\mathcal{K}) = O(\mathcal{K}) \ltimes \mathcal{K}$, define an embedding from $\mathrm{Isom}(\mathcal{K})$ to $O(1, \infty)$ by

$$(A, \xi) \mapsto \begin{pmatrix} 1 + \frac{1}{2}\langle \xi, \xi \rangle & \frac{1}{2}\langle \xi, \xi \rangle & -(A\xi)^t \\ -\frac{1}{2}\langle \xi, \xi \rangle & 1 - \frac{1}{2}\langle \xi, \xi \rangle & (A\xi)^t \\ -\xi & -\xi & A \end{pmatrix},$$

where $A \in O(\mathcal{K})$ and $\xi \in \mathcal{K}$.

Condition (2) implies condition (3) immediately, because the embedding of $\mathcal{H}$ into $\mathcal{H}_\mathbb{C}$ induces an embedding of $O(1, \infty)$ into $U(1, \infty)$.

Finally, we show that condition (3) implies condition (1). Denote by $d(\cdot, \cdot)$ the distance function on $\mathbb{H}^\infty(\mathbb{C})$. By [FH74, Corollary 8.2], the kernel $\log \cosh d$ is conditionally negative definite on $\mathbb{H}^\infty(\mathbb{C})$. Fix $x_0 \in \mathbb{H}^\infty(\mathbb{C})$. If G is a group acting isometrically and metrically properly on $\mathbb{H}^\infty(\mathbb{C})$, then the function $g \mapsto \log \cosh d(gx_0, x_0)$ is conditionally negative definite and proper on G. So G has the Haagerup property.   □

It is interesting to speculate whether there are other geometric characterizations of the Haagerup property.

7.5   Other dynamical characterizations

7.5.1   Actions on infinite measure spaces

In [RS98], Robertson and Steger obtain the following characterization of property (T) for countable groups: a countable group Γ has property (T) if and only if, for every measure-preserving action of Γ on a measure space $(\Omega, \mathcal{B}, \mu)$, and every set $S \in \mathcal{B}$ such that $\mu(S \triangle gS) < +\infty$ for every $g \in \Gamma$, one has $\sup_{g \in \Gamma} \mu(S \triangle gS) < +\infty$. We do not know whether this can be generalized to locally compact, nondiscrete groups.

For countable groups, we have the parallel characterization of the Haagerup property.

Proposition 7.5.1. *Let Γ be a countable group. Then Γ has the Haagerup property if and only if there exists a measure-preserving action of Γ on a measure space $(\Omega, \mathcal{B}, \mu)$, and a measurable set S such that $\mu(S \triangle gS) < +\infty$ for every $g \in \Gamma$, and the function $g \mapsto \mu(S \triangle gS)$ is proper on Γ.*

Proof. Assume first that Γ has the Haagerup property. Write Ω for the set

$$\{0,1\}^{\Gamma} \setminus \{(\ldots,0,0,0,\ldots),(\ldots,1,1,1,\ldots)\}.$$

Let ψ be a proper, conditionally negative definite function on Γ. By [RS98, Prop 1.4], there exists a regular Borel measure μ on Ω such that

$$\mu(S_g \triangle S_h) = \sqrt{\psi(g^{-1}h)},$$

where $S_g = \{x \in \Omega : x_g = 1\}$ for all $g \in \Gamma$. Moreover, the proof of [RS98, Thm 2.1] shows that μ is Γ-invariant. Hence

$$h \mapsto \sqrt{\psi(h)} = \mu(S_1 \triangle S_h) = \mu(S_1 \triangle h S_1)$$

is a proper function on Γ.

Now we prove the converse. Denote by χ_B the characteristic function of B in $\mathcal{B}$. Let π be the natural representation of Γ on $L^2(\Omega, \mu)$, and let X be the Γ-orbit of S in $\mathcal{B}$. The assumption implies that, for $S_1, S_2 \in X$, the function $\chi_{S_1} - \chi_{S_2}$ is in $L^2(\Omega, \mu)$. Set then

$$c(S_1, S_2) = \chi_{S_1} - \chi_{S_2} \quad \text{and} \quad d(S_1, S_2) = \|\chi_{S_1} - \chi_{S_2}\|.$$

This defines a Γ-equivariant cocycle on X taking values in $L^2(\Omega, \mu)$, and a distance function on X for which Γ acts isometrically. Then Γ acts properly on X, since

$$d(gS, S)^2 = \mu(S \triangle gS) \quad \forall g \in \Gamma,$$

From Proposition 7.4.1, Γ has the Haagerup property. $\square$

Again, we do not know whether this can be generalized to nondiscrete groups.

7.5.2 Invariant probability measures

Let G be a second countable, locally compact group acting by homeomorphisms on a compact metrizable space X. We denote by $M(X)$ the convex set of probability measures on X, equipped with the weak-*topology, and by $M_G(X)$ the closed convex subset of G-invariant measures. The extreme points of $M_G(X)$ are the *ergodic* probability measures on X.

Denote by G^+ the one-point compactification $G \cup \{\infty\}$ of G. Let Σ be the set of closed subsets in G^+ containing ∞; equipped with the Hausdorff topology, this is a metrizable compact space, on which G acts by left translations. In [GW97], E. Glasner and B. Weiss obtain a remarkable new characterization of property (T) for a second countable, locally compact group G, by proving that the following statements are equivalent:

(1) G has property (T);
(2) for every action of G by homeomorphisms on a compact, metrizable space X with $M_G(X)$ nonempty, the set of ergodic measures is closed in $M_G(X)$;
(3) the set of ergodic measures is closed in $M_G(\Sigma)$;
(4) the set of ergodic measures is not dense in $M_G(\Sigma)$.

According to the philosophy that, to any characterization of property (T) there is a parallel characterization of the Haagerup property, there should be a definition of the Haagerup property corresponding to the above definition of property (T). What is it?

Bibliography

[AC83]   H. Araki and M. Choda. Property T and actions on the hyperfinite
           II$_1$-factor. *Math. Japon.*, 28(2):205–209, 1983.

[AD95]   C. Anantharaman-Delaroche. Amenable correspondences and ap-
           proximation properties for von Neumann algebras. *Pacific J.
           Math.*, 171(2):309–341, 1995.

[ADY96]   J.-P. Anker, E. Damek, and C. Yacoub. Spherical analysis on har-
           monic AN groups. *Ann. Scuola Norm. Sup. Pisa Cl. Sci. (4)*,
           23(4):643–679, 1996.

[AEG94]   S. Adams, G.A. Elliott, and T. Giordano. Amenable actions of
           groups. *Trans. Amer. Math. Soc.*, 344(2):803–822, 1994.

[Alp82]   R. Alperin. Locally compact groups acting on trees and property
           T. *Monatsh. Math.*, 93(4):261–265, 1982.

[AW81]   C.A. Akemann and M. Walter. Unbounded negative definite func-
           tions. *Canad. J. Math.*, 33(4):862–871, 1981.

[Bal95]   W. Ballmann. *Lectures on spaces of nonpositive curvature*. DMV
           Seminar, Band 25. Birkhäuser Verlag, Basel, 1995.

[Bau71]   G. Baumslag. Positive one-relator groups. *Trans. Amer. Math.
           Soc.*, 156:165–183, 1971.

[Bau93]   G. Baumslag. *Topics in combinatorial group theory*. Birkhäuser
           Verlag, Basel, 1993.

[BB95]   W. Ballmann and M. Brin. Orbihedra of nonpositive curvature.
           Inst. Hautes Études Sci. Publ. Math., 82:169–209, 1995.

[BBV99]   C. Béguin, H. Bettaieb, and A. Valette. K-theory for C^*-algebras
           of one-relator groups. *K-Theory*, 16(3):277–298, 1999.

[BCH94]   P. Baum, A. Connes, and N. Higson. Classifying space for proper
           actions and K-theory of group C^*-algebras. In *C^*-algebras: 1943–
           1993 (San Antonio, TX, 1993)*, volume 167 of *Contemporary Math-
           ematics*, pages 240–291. Amer. Math. Soc., Providence, RI, 1994.

© Springer Basel 2001
P.-A. Cherix et al., *Groups with the Haagerup Property*,
Modern Birkhäuser Classics, DOI 10.1007/978-3-0348-0906-1

[BCS00]   C. Béguin and T. Ceccherini-Silberstein. Formes faibles de moyen-nabilité pour les groupes à un relateur. *Bull. Belg. Math. Soc. Simon Stevin*, 7(1):135–148, 2000.

[BCV95]   M.E.B. Bekka, P.-A. Cherix, and A. Valette. Proper affine isometric actions of amenable groups. In *Novikov conjectures, index theorems and rigidity, Vol. 2 (Oberwolfach, 1993)*, volume 227 of *London Math. Soc. Lecture Notes*, pages 1–4. Cambridge Univ. Press, Cambridge, 1995.

[BJS88]   M. Bożejko, T. Januszkiewicz, and R. J. Spatzier. Infinite Coxeter groups do not have Kazhdan's property. *J. Operator Theory*, 19(1):63–67, 1988.

[BM97]   M. Burger and S. Mozes. Finitely presented simple groups and products of trees. *C. R. Acad. Sci. Paris Sér. I Math.*, 324(7):747–752, 1997.

[Boc91]   F. Boca. Free products of completely positive maps and spectral sets. *J. Funct. Anal.*, 97(2):251–263, 1991.

[Boż89]   M. Bożejko. Positive-definite kernels, length functions on groups and a noncommutative von Neumann inequality. *Studia Math.*, 95(2):107–118, 1989.

[BR88]   V. Bergelson and J. Rosenblatt. Mixing actions of groups. *Illinois J. Math.*, 32(1):65–80, 1988.

[BR95]   T. Bates and A.G. Robertson. Positive definite functions and relative property (T) for subgroups of discrete groups. *Bull. Austral. Math. Soc.*, 52(1):31–39, 1995.

[Bri]   M. Bridson. Remarks on a talk of Higson. 3 June 1997.

[BS62]   G. Baumslag and D. Solitar. Some two-generator one-relator non-Hopfian groups. *Bull. Amer. Math. Soc.*, 68:199–201, 1962.

[BŚ97]   W. Ballmann and J. Świątkowski. On L^2-cohomology and property (T) for automorphism groups of polyhedral cell complexes. *Geom. Funct. Anal.*, 7(4):615–645, 1997.

[BT79]   L. Baggett and K. Taylor. A sufficient condition for the complete reducibility of the regular representation. *J. Funct. Anal.*, 34(2):250–265, 1979.

[BTV95]   J. Berndt, F. Tricerri, and L. Vanhecke. *Generalized Heisenberg groups and Damek–Ricci harmonic spaces*, volume 1598 of *Lecture Notes in Mathematics*. Springer Verlag, Berlin, Heidelberg, New York, 1995.

[CDKR91]   M. Cowling, A.H. Dooley, A. Korányi, and F. Ricci. H-type groups and Iwasawa decompositions. *Adv. Math.*, 87(1):1–41, 1991.

[CDKR98] M. Cowling, A.H. Dooley, A. Korányi, and F. Ricci. An approach to symmetric spaces of rank one via groups of Heisenberg type. *J. Geom. Anal.*, 8(2):199–237, 1998.

[CH89] M. Cowling and U. Haagerup. Completely bounded multipliers of the Fourier algebra of a simple Lie group of real rank one. *Invent. Math.*, 96(3):507–549, 1989.

[Cho83] M. Choda. Group factors of the Haagerup type. *Proc. Japan Acad. Ser. A Math. Sci.*, 59(5):174–177, 1983.

[CK84] M. Cowling and A. Korányi. Harmonic analysis on Heisenberg type groups from a geometric viewpoint. In *Lie group representations, III (College Park, Md., 1982/1983)*, volume 1077 of *Lecture Notes in Mathematics*, pages 60–100. Springer Verlag, Berlin, Heidelberg, New York, 1984.

[Con75] A. Connes. Outer conjugacy classes of automorphisms of factors. *Ann. Sci. École Norm. Sup. (4)*, 8(3):383–419, 1975.

[Cow83] M. Cowling. Harmonic analysis on some nilpotent Lie groups (with application to the representation theory of some semisimple Lie groups). In *Topics in modern harmonic analysis, Vol. I, II (Turin/Milan, 1982)*, pages 81–123. Istituto Nazionale di Alta Matematica Francesco Severi, Rome, 1983.

[Cow79] M. Cowling. Sur les coefficients des représentations unitaires des groupes de Lie simples. In *Analyse harmonique sur les groupes de Lie II. (Séminaire Nancy-Strasbourg 1976–1978)*, volume 739 of *Lecture Notes in Mathematics*, pages 132–178. Springer Verlag, 1979.

[Cun83] J. Cuntz. *K*-theoretic amenability for discrete groups. *J. reine angew. Math.*, 344:180–195, 1983.

[CW80] A. Connes and B. Weiss. Property T and asymptotically invariant sequences. *Israel J. Math.*, 37(3):209–210, 1980.

[Dam87a] E. Damek. Curvature of a semidirect extension of a Heisenberg type nilpotent group. *Colloq. Math.*, 53(2):249–253, 1987.

[Dam87b] E. Damek. The geometry of a semidirect extension of a Heisenberg type nilpotent group. *Colloq. Math.*, 53(2):255–268, 1987.

[DB97a] B. Di Blasio. An extension of the theory of Gelfand pairs to radial functions on Lie groups. *Boll. Un. Mat. Ital. B (7)*, 11(3):623–642, 1997.

[DB97b] B. Di Blasio. Positive definite spherical functions on harmonic space *NA*. *Boll. Un. Mat. Ital. A (7)*, 11(3):759–767, 1997.

[DCH85]   J. De Cannière and U. Haagerup. Multipliers of the Fourier alge-
          bras of some simple Lie groups and their discrete subgroups. *Amer.
          J. Math.*, 107(2):455–500, 1985.

[DR92]   E. Damek and F. Ricci. Harmonic analysis on solvable extensions
          of H-type groups. *J. Geom. Anal.*, 2(3):213–248, 1992.

[DZ99]   A.H. Dooley and G.K. Zhang. Spherical functions on harmonic
          extensions of H-type groups. *J. Geom. Anal.*, 9(2):247–255, 1999.

[Eym64]   P. Eymard. L'algèbre de Fourier d'un groupe localement compact.
          Bull. Soc. Math. France, 92:181–236, 1964.

[Eym72]   P. Eymard. *Moyennes invariantes et représentations unitaires*,
          volume 300 of *Lecture Notes in Mathematics*. Springer Verlag,
          Berlin, Heidelberg, New York, 1972.

[Far00]   D.S. Farley. *Finiteness in $CAT(0)$ properties of diagram groups.*
          PhD thesis, SUNY Binghampton, May 2000.

[FH74]   J. Faraut and K. Harzallah. Distances hilbertiennes invariantes sur
          un espace homogène. *Ann. Inst. Fourier (Grenoble)*, 24(3):171–
          217, 1974.

[FJ77]   M. Flensted-Jensen. Spherical functions on a simply connected
          semisimple Lie group. II. The Paley-Wiener theorem for the rank
          one case. *Math. Ann.*, 228(1):65–92, 1977.

[Fol75]   G. B. Folland. Subelliptic estimates and function spaces on nilpo-
          tent Lie groups. *Ark. Mat.*, 13(2):161–207, 1975.

[Gal]   S.R. Gal. Property of Haagerup for amalgams. In preparation.

[GJ00]   S.R. Gal and T. Januszkiewicz. New a-T-menable HNN-extensions.
          Preprint, January 2000.

[Got50]   M. Gotô. Faithful representations of Lie groups. II. *Nagoya Math.
          J.*, 1:91–107, 1950.

[Gro87]   M. Gromov. Hyperbolic groups. In S.M. Gersten, editor, *Essays
          in group theory*, volume 8 of *M.S.R.I. Publications*, pages 75–263.
          Springer Verlag, Berlin, Heidelberg, New York, 1987.

[Gro88]   M. Gromov. Rigid transformations groups. In D. Bernard and
          Y. Choquet Bruhat, editors, *Géométrie différentielle, variétés
          complexes, feuilletages riemanniens (Paris, 1986)*, volume 33 of
          Travaux en cours, pages 65–139. Hermann, Paris, 1988.

[Gro93]   M. Gromov. Asymptotic invariants of infinite groups. In G.A.
          Niblo and M.A. Roller, editors, *Geometric group theory, Vol. 2
          (Sussex, 1991)*, volume 182 of *London Math. Soc. Lecture Notes*,
          pages 1–295. Cambridge Univ. Press, Cambridge, 1993.

[Gro99]   M. Gromov. Spaces and questions. Preprint, November 1999.

[GW97]   E. Glasner and B. Weiss. Kazhdan's property T and the geome-
           try of the collection of invariant measures. *Geom. Funct. Anal.*,
           7(5):917–935, 1997.

[Haa79]   U. Haagerup. An example of a nonnuclear C^*-algebra, which has
           the metric approximation property. *Invent. Math.*, 50(3):279–293,
           1979.

[Hem76]   J. Hempel. *3-Manifolds*, volume 86 of *Annals of Math. Studies*.
           Princeton University Press, Princeton, N. J., 1976.

[Hig00]   N. Higson. Bivariant K-theory and the Novikov conjecture. Geom.
           Funct. Anal. 10 (2000), 563–581.

[HJ]   P. de la Harpe and V. Jones. An introduction to C*-algebras.
           Université de Genève, 1995.

[HK97]   N. Higson and G. Kasparov. Operator K-theory for groups which
           act properly and isometrically on Hilbert space. *Electron. Res.
           Announc. Amer. Math. Soc.*, 3:131–142 (electronic), 1997.

[HM79]   R.E. Howe and C.C. Moore. Asymptotic properties of unitary
           representations. *J. Funct. Anal.*, 32(1):72–96, 1979.

[Hoc65]   G. Hochschild. *The structure of Lie groups*. Holden-Day Inc., San
           Francisco, 1965.

[HP98]   F. Haglund and F. Paulin. Simplicité de groupes d'automorphismes
           d'espaces à courbure négative. In *The Epstein birthday Schrift*,
           pages 181–248 (electronic). Geom. Topol., Coventry, 1998.

[HR00]   N. Higson and J. Roe. Amenable group actions and the Novikov
           conjecture. *J. reine angew. Math.*, 519:143–153, 2000.

[HV89]   P. de la Harpe and A. Valette. *La propriété (T) de Kazhdan pour les
           groupes localement compacts (avec un appendice de Marc Burger)*,
           volume 175 of *Astérisque*. Société Mathématique de France, Paris,
           1989.

[IN96]   A. Iozzi and A. Nevo. Algebraic hulls and the Følner property.
           Geom. Funct. Anal., 6(4):666–688, 1996.

[Jan98]   T. Januszkiewicz. For Coxeter groups $z^{|g|}$ is a coefficient of a
           uniformly bounded representation. Preprint, December 1998.

[Jol93]   P. Jolissaint. Property T for discrete groups in terms of their
           regular representation. *Math. Ann.*, 297(3):539–551, 1993.

[Jol00]   P. Jolissaint. Borel cocycles, approximation properties and relative
           property T. *Ergodic Theory Dynam. Systems*, 20(2):483–499, 2000.

[Jon83]   V.F.R. Jones. A converse to Ocneanu's theorem. *J. Operator
           Theory*, 10(1):61–63, 1983.

[Jul98]   P. Julg. Travaux de N. Higson et G. Kasparov sur la conjecture de Baum-Connes. Exp. no. 841. In *Séminaire Bourbaki 1997/98*, volume 252 of *Astérisque*, pages 151–183. Société Mathématique de France, Paris, 1998.

[JV84]   P. Julg and A. Valette. K-theoretic amenability for $SL_2(\mathbf{Q}_p)$, and the action on the associated tree. *J. Funct. Anal.*, 58(2):194–215, 1984.

[JV91]   P. Jolissaint and A. Valette. Normes de Sobolev et convoluteurs bornés sur $L^2(G)$. *Ann. Inst. Fourier (Grenoble)*, 41(4):797–822, 1991.

[JW77]   K.D. Johnson and N.R. Wallach. Composition series and inter-twining operators for the spherical principal series. I. *Trans. Amer. Math. Soc.*, 229:137–173, 1977.

[Kap80]   A. Kaplan. Fundamental solutions for a class of hypoelliptic PDE generated by composition of quadratic forms. *Trans. Amer. Math. Soc.*, 258(1):147–153, 1980.

[Keh84]   E.T. Kehlet. Cross sections for quotient maps of locally compact groups. *Math. Scand.*, 55(1):152–160, 1984.

[Kor82]   A. Korányi. Some applications of Gel'fand pairs in classical analysis. In *Harmonic analysis and group representations*, pages 333–348. Liguori, Naples, 1982.

[Kos69]   B. Kostant. On the existence and irreducibility of certain series of representations. *Bull. Amer. Math. Soc.*, 75:627–642, 1969.

[Lep68]   H. Leptin. Sur l'algèbre de Fourier d'un groupe localement compact. *C. R. Acad. Sci. Paris Sér. A-B*, 266:A1180–A1182, 1968.

[LH90]   M. Lemvig Hansen. Weak amenability of the universal covering group of SU(1, n). *Math. Ann.*, 288(3):445–472, 1990.

[LM92]   A. Lubotzky and S. Mozes. Asymptotic properties of unitary representations of tree automorphisms. In M.A. Picardello, editor, *Harmonic analysis and discrete potential theory (Frascati, 1991)*, pages 289–298. Plenum, New York, 1992.

[LS77]   R.C. Lyndon and P.E. Schupp. *Combinatorial group theory*. Ergebnisse der Mathematik und ihrer Grenzgebiete, Band 89. Springer Verlag, Berlin, Heidelberg, New York, 1977.

[Mar91]   G.A. Margulis. *Discrete subgroups of semisimple Lie groups*. Ergebnisse der Mathematik und ihrer Grenzgebiete. 3 Folge, Band 17. Springer Verlag, Berlin, Heidelberg, New York, 1991.

[Mas67]   W.S. Massey. *Algebraic topology: An introduction*. Harcourt, Brace & World, Inc., New York, 1967.

[Ner98]   Y.A. Neretin.   Notes on affine isometric actions of discrete groups. In H. Meyer and J. Marion, editors, *Analysis on infinite-dimensional Lie groups and algebras (Marseille, 1997)*, pages 274–320. World Sci. Publishing, River Edge, NJ, 1998.

[NR97]   G. Niblo and L. Reeves. Groups acting on CAT(0) cube complexes. *Geom. Topol.*, 1:1–7 (electronic), 1997.

[Ocn85]   A. Ocneanu. *Actions of discrete amenable groups on von Neumann algebras*, volume 1138 of *Lecture Notes in Mathematics*. Springer Verlag, Berlin, Heidelberg, New York, 1985.

[OO98]   H. Oyono-Oyono. La conjecture de Baum-Connes pour les groupes agissant sur les arbres. *C. R. Acad. Sci. Paris Sér. I Math.*, 326(7):799–804, 1998.

[Pas77]   D.S. Passman. *The algebraic structure of group rings*. Pure and Applied Mathematics. Wiley-Interscience [John Wiley & Sons], New York, 1977.

[Pav91]   M. Pavone. On the geodesic distance and group actions on trees. *Math. Proc. Cambridge Philos. Soc.*, 110(1):67–70, 1991.

[Ped79]   G.K. Pedersen. C^*-*algebras and their automorphism groups*. Academic Press Inc. [Harcourt Brace Jovanovich Publishers], London, 1979.

[Ric85]   F. Ricci. Commutative algebras of invariant functions on groups of Heisenberg type. *J. London Math. Soc. (2)*, 32(2):265–271, 1985.

[Rie82]   C. Riehm. The automorphism group of a composition of quadratic forms. *Trans. Amer. Math. Soc.*, 269(2):403–414, 1982.

[Rin88]   H. Rindler. Groups of measure preserving transformations. *Math. Ann.*, 279(3):403–412, 1988.

[Rob93]   A.G. Robertson. Property (T) for II_1 factors and unitary representations of Kazhdan groups. *Math. Ann.*, 296(3):547–555, 1993.

[Rob98]   A.G. Robertson. Crofton formulae and geodesic distance in hyperbolic spaces. *J. Lie Theory*, 8(1):163–172, 1998.

[Ros81]   J. Rosenblatt. Uniqueness of invariant means for measure-preserving transformations. *Trans. Amer. Math. Soc.*, 265(2):623–636, 1981.

[RS98]   A.G. Robertson and T. Steger. Negative definite kernels and a dynamical characterization of property (T) for countable groups. *Ergodic Theory Dynam. Systems*, 18(1):247–253, 1998.

[Saa96]   L. Saal. The automorphism group of a Lie algebra of Heisenberg type. *Rend. Sem. Mat. Univ. Politec. Torino*, 54(2):101–113, 1996.

[Sag95]   M. Sageev. Ends of group pairs and non-positively curved cube complexes. *Proc. London Math. Soc. (3)*, 71(3):585–617, 1995.

[Sch80]   K. Schmidt. Asymptotically invariant sequences and an action of
          SL(2, **Z**) on the 2-sphere. *Israel J. Math.*, 37(3):193–208, 1980.

[Sch81]   K. Schmidt. Amenability, Kazhdan's property T, strong ergodicity
          and invariant means for ergodic group-actions. *Ergodic Theory
          Dynamical Systems*, 1(2):223–236, 1981.

[Sch96]   K. Schmidt. From infinitely divisible representations to cohomo-
          logical rigidity. In *Analysis, geometry and probability*, Texts Read.
          Math., pages 173–197. Hindustan Book Agency, Delhi, 1996.

[Ser77]   J.-P. Serre. *Arbres, amalgames*, SL$_2$, volume 46 of *Astérisque*.
          Société Mathématique de France, Paris, 1977.

[Sha99]   Y. Shalom. Rigidity, unitary representations of semisimple groups,
          and fundamental groups of manifolds with rank one transformation
          group. Preprint, 1999.

[Szw91]   R. Szwarc. Groups acting on trees and approximation properties
          of the Fourier algebra. *J. Funct. Anal.*, 95(2):320–343, 1991.

[Thu82]   W.P. Thurston. Three-dimensional manifolds, Kleinian groups and
          hyperbolic geometry. *Bull. Amer. Math. Soc. (N.S.)*, 6(3):357–381,
          1982.

[Tu99]   J.-L. Tu. La conjecture de Baum-Connes pour les feuilletages
          moyennables. *K-Theory*, 17(3):215–264, 1999.

[Val89]   A. Valette. The conjecture of idempotents: a survey of the C^*-
          algebraic approach. *Bull. Soc. Math. Belg. Sér. A*, 41(3):485–521,
          1989.

[Val90]   A. Valette. Les représentations uniformément bornées associées à
          un arbre réel. *Bull. Soc. Math. Belg. Sér. A*, 42(3):747–760, 1990.

[Val93]   A. Valette. Weak amenability of right-angled Coxeter groups. *Proc.
          Amer. Math. Soc.*, 119(4):1331–1334, 1993.

[Val94]   A. Valette. Old and new about Kazhdan's property (T). In V.
          Baldoni and M. Picardello, editors, *Representations of Lie groups
          and quantum groups (Trento, 1993)*, Pitman Res. Notes in Math.,
          pages 271–333. Longman Sci. Tech., Harlow, 1994.

[VGG73]   A.M. Vershik, I.M. Gel'fand, and M.I. Graev. Representations of
          SL(2,R) where R is a ring of functions. *Russian Math. Surveys*,
          285:87–132, 1973.

[VGG74]   A.M. Vershik, I.M. Gel'fand, and M.I. Graev. Irreducible represen-
          tations of the group G^X and cohomology. *Funkt. Anal. i Prilozh.*,
          8:67–69, 1974.

[Vig80]   M.-F. Vignéras. *Arithmétique des algèbres de quaternions*, volume
          800 of *Lecture Notes in Math.* Springer Verlag, Berlin, Heidelberg,
          New York, 1980.

[War72]   G. Warner. *Harmonic analysis on semi-simple Lie groups. I.* Die Grundlehren der mathematischen Wissenschaften, Band 188. Springer Verlag, Berlin, Heidelberg, New York, 1972.

[Wat81]   Y. Watatani. Property (T) of Kazhdan implies property (FA) of Serre. *Math. Japon.*, 27:97–103, 1981.

[Yu00]   G.-L. Yu. The coarse Baum-Connes conjecture for spaces which admit a uniform embedding into Hilbert space. *Invent. Math.*, 139(1):201–240, 2000.

[Zim84a] R.J. Zimmer. *Ergodic theory and semisimple groups.* Birkhäuser Verlag, Basel, 1984.

[Zim84b] R.J. Zimmer. Kazhdan groups acting on compact manifolds. *Invent. Math.*, 75(3):425–436, 1984.

Index

(k, l)-complex, 5
s-density, 38
C*-algebra of a group
  maximal or full, 7
  reduced, 7

a-T-menability, 1
a-T-menable, 1
action
  centrally free, 30
  essentially free, 22
  strongly ergodic, 18
  strongly mixing, 18
    on a von Neumann algebra, 20
amalgamated product, 93
  algebraic, 100
approximate unit, 7
asymptotically invariant sequence, 18
  in a von Neumann algebra, 20
  nontrivial, 18
automorphism
  centrally trivial, 30

Bass–Serre theory, 97

CAR algebra, 25
Chasles' relation, 109
cocycle
  G-equivariant, 35
  associated with π, 17
  Busemann, 34

with value in $\mathcal{H}$, 92
complete subset, 100
conjecture
  Baum–Connes, 8
  idempotents, 8
  Kaplansky–Kadison, 8
  Novikov, 9
Cowling–Haagerup constant, 7

factor
  hyperfinite, 20
  McDuff, 31
  Powers, 20
Fourier algebra of a group, 7
function
  Bruhat, 87
  conditionally negative definite, 15
  hypergeometric, 58
  positive definite
    normalized, 1
  spherical, 58
Følner sequence
  in a probability space, 18
  in a von Neumann algebra, 20

Gel'fand–Naimark–Segal
  construction, 6
group
  adèle, 86
  Baumslag–Solitar monster, 106
  braid, 107
  Heisenberg, 45

© Springer Basel 2001
P.-A. Cherix et al., *Groups with the Haagerup Property*,
Modern Birkhäuser Classics, DOI 10.1007/978-3-0348-0906-1